Electronic Circuits

Volume 1.0

Disclaimer

The circuits, software or related documentation in this book are NOT designed nor intended for use (whether free or sold) as on-line control equipment in hazardous environments requiring fail-safe performance, such as, but not limited to, in the operation of nuclear facilities, aircraft navigation or communication systems, air traffic control, direct life support machines or weapons systems in which the failure of the hardware or software could lead directly to death, personal injury, or severe physical or environmental damage ("high risk activities")

The author(s) and publisher(s) take no responsibility for damages or injuries of any kind that may arise from the use or misuse of the circuits and/or software in this collection.

The author(s) and publisher(s) specifically disclaim any express or implied warranty or fitness for high risk activities. The circuits, software and related documentation are without warranty of any kind. The author(s) and publisher(s) expressly disclaim all other warranties, express or implied, including, but not limited to, the implied warranties of merchantability and fitness for a particular purpose. Under no circumstances shall the author(s) and publisher(s) be liable for any incidental, special or consequential damages that result from the use or inability to use the circuits and software or related documentation, even if he has been advised of the possibility of such damages.

Electronic Circuits

Volume 1.0

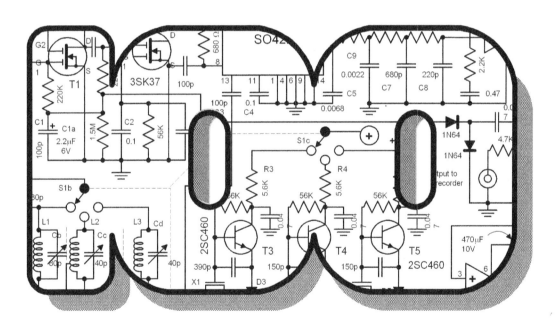

www.intellin.org

Published by the

 Intellin Organization

Copyright 2006 by
Intellin Organization LLC
www.intellin.org

First year of publication 2006

All rights reserved.
No part of this book may be
reproduced in any form or by
any means, except brief quotations
for a review, without permission
in writing from the author(s).

Disclaimer:

The circuits, software or related documentation in this book are NOT designed nor intended for use (whether free or sold) as on-line control equipment in hazardous environments requiring fail-safe performance, such as, but not limited to, in the operation of nuclear facilities, aircraft navigation or communication systems, air traffic control, direct life support machines or weapons systems in which the failure of the hardware or software could lead directly to death, personal injury, or severe physical or environmental damage ("high risk activities")

The author(s) and publisher(s) take no responsibility for damages or injuries of any kind that may arise from the use or misuse of the circuits in this collection.

The author(s) and publisher(s) specifically disclaim any express or implied warranty or fitness for high risk activities. The circuits, software and related documentation are without warranty of any kind. The author(s) and publisher(s) expressly disclaim all other warranties, express or implied, including, but not limited to, the implied warranties of merchantability and fitness for a particular purpose. Under no circumstances shall the author(s) and publisher(s) be liable for any incidental, special or consequential damages that result from the use or inability to use the circuits and software or related documentation, even if he has been advised of the possibility of such damages.

ISBN   1-4196-4399-1

PREFACE

Congratulations for having the first volume of ready-to-apply circuits. You got the luxury of being able to design and assemble electronic modules fast and worry free. It's a sure way to optimize satisfaction in your hobby. If you are a professional electronic designer, it will help you beat the competition. Speed, efficiency, short development periods, error-free, user and maintenance friendly: these are the factors critical for success. This invaluable book filled with **100** practical ideas will help you beat project deadlines. Make your ideas work!

Make your creativity pay! And all that JUST IN TIME!

informative...
practical...
professional...
versatile...

Acknowledgments

Many Thanks to...

Engineer Mischa (Optical Recognition)
Engineer Salinger (Electronics)
Engineer P. Schmidt (Cybernetics)
Engineer N. Lay (Robotics)

INTRODUCTION

This collection contains 100 practical electronic circuits grouped in ten general applications. Since most of the circuits are not limited to a single application, a circuit may have found its way into another group. This is one proof of the versatility of the circuits. Creativity needs versatility . You can combine several circuits into one large module to create a powerful electronic device especially designed for your exclusive project.

The table of contents lists the groups and titles of the circuits. The page number where a group begins can be found in this table. To find a particular circuit, turn to the group's beginning page. On this page, the circuits are again listed with their page numbers. To quickly find an application group, use the black markers on the edge of the pages. These markers coincide with the markers in the table of contents.

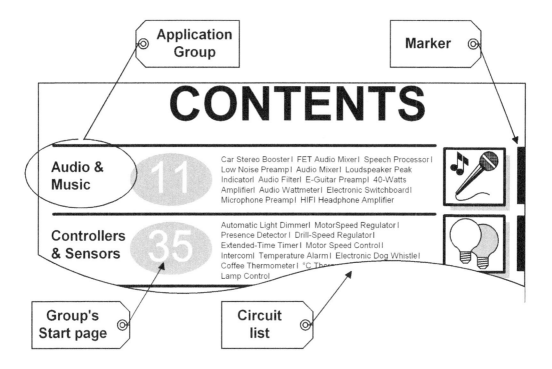

The transistors used in the circuits have more than one possible replacements. Their pin designations are also shown in details. This feature can help avoid unnecessary delays. The pins shown are either in the bottom view or front view of the transistor unless otherwise noted. Large transistors that cannot or are not planned to be installed directly on the PCB must be installed on a heatsink. A dashed circle around a transistor means the transistor must be heatsinked.

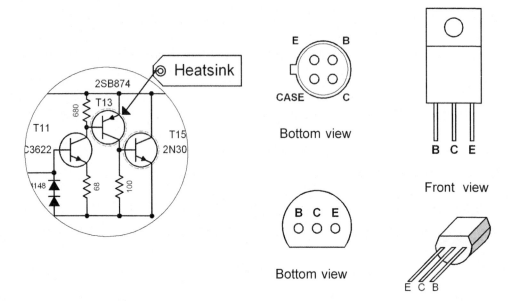

The capacitor values are given in microfarad unless otherwise specified. Electrolytic or polarized capacitors are marked with a plus sign in the diagram. This plus sign coincides with the capacitor's positive polarity in the circuit. Additionally, their voltage ratings are also given. Nonpolar capacitors are ceramic types and rated with 50 volts.

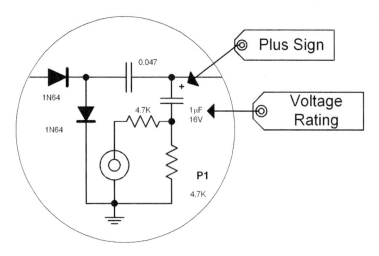

The resistor values are given in ohms (W), rated 1/4 watts and are of carbon film type unless otherwise specified.

CONTENTS

Audio & Music		Single IC 2.5W Amplifier-Audio Peak Meter-Very Low Noise Mic Amp-Signal Clip Indicator-Dyna Audio Compressor-Stereo Audio Mixer-Tunable Filter Circuit-Channel Balance Indicator-Tweeter Guardian-Low Noise Preamp-Envelope Sampler-Mic Processor	
Hobby & Shop		Digital Bandpass Filter-Transistor Solar Cells- Hybrid Cascaded Transistor-Adjustable Dummy Load-Optocoupler-Automatic Battery Charger-Voltage Frequency Converter-Music in a chip-Ultrasound Receiver	
House & Car		Atmospheric Disturbance Detector-Electronic Fuse-Lamp Dimmer-Intercom-Optically Coded Key-Thermometer-Automatic Lamp Regulator-Universal Timer-Smoke Alarm-Electronic Selector-Infrared Switch-DC Flourescent Lamp-Thermonitor-Live Wire Sensor	
Radio Frequency		Diode AM Receiver-SSB from SW Adapter-VHF Dip Meter-Morse Code Filter-VLF Converter-Active Impedance Converter-VFO Stabilizer Circuit-	
Hobby & Games		Voice Operated Switch-Soldering Iron Regulator-Electronic Pool-Alternating Lamps-Running Light-Projector Film Changer-Auto Soldering-Bipolar Stepmotor Controller	
Power Supplies & Chargers		Polarity Protected Charger-Overvoltage Crowbar-Power Supply Regulator-PS with Dissipation Limiter-Stable Z-voltage Source-DC to DC Converter-Versatile Power Supply-Symmetrical Auxiliary PS-Low Drop Regulator	
Testers & Meters		Diode Tester-Logic Probe-Poorman's Logic Analyzer-C° to Frequency Converter-Radio Meter-AF Counters-LED Constant Current Source- Wideband Signal Injector-Tendency Indicator-Wienbridge Oscillator-Cheap Frequency Counter-Light-Frequency Converter-BW TV Pattern Generator-Acoustic Continuity Tester	
Digital & Computers		Tape Content Monitor-Infrared Interface Circuit-Two way RS 232-Flip-Flop from Inverters-Hardware Screensaver-Monitor Driver Circuit	
Oscillators & Counters		Function Generator-Duty Cycle Generator-Start-Stop Generator-Crystal Controlled Timebase-48-MHz Clock Generator-Sine to Square/Trianglewave	
Auxiliary		Light Activated Switch-Sound Generator-Automatic Resetter-DC Voltage Double-Synchronized Sawtooth-Running Light-Liner Optocoupler-Adjustable Zener Diode-Signal Light Clicker-Headlamp Dimmer-1 Chip TV Audio-Simple Electronic Organ-Voltage/Frequency Converter-Debounced Pulse Generator	

Electronic Circuits - 1.0

This page is intentionally blank.

AUDIO & MUSIC

- **12** Single IC 2.5W Amplifier
- **14** Audio Peakmeter
- **16** Very Low Noise Mic Amp
- **18** Signal Clip Indicator
- **20** Dyna Audio Compressor
- **22** Stereo Audio Mixer
- **24** Tunable Filter Circuit
- **25** Channel Balance Indicator
- **27** Tweeter Guardian
- **29** Low Noise Preamp
- **31** Envelope Sampler
- **32** Microphone Processor
- **33** HIFI Audio Mixer Module

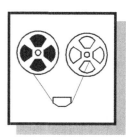

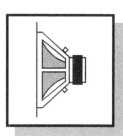

Electronic Circuits - 1.0

1. Single IC 2.5W Amplifier

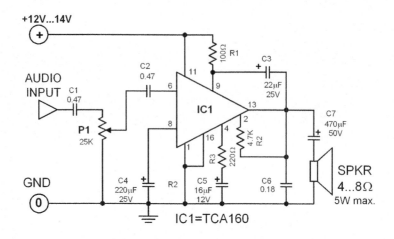

Diagram 1.0 Single IC 2.5W Amplifier

There are many cases where you desperately need a simple to build and inexpensive amplifier that delivers moderate power. Here it is. This compact amplifier delivers just enough power output so that you can hear the audio signal coming from any device. It delivers up to 2.5 watts audio output.

Formula for P
$P = \dfrac{Vb2}{8RI}$
where Vb= actual supply voltage and RI= actual speaker impedance.

Audio & Music

The heart of the amplifier is a single compact IC which makes the circuit simple to build and eventually troubleshoot. The IC used in this circuit is a TCA160 which integrates a full-amplifier circuit including preamp and driver stages. The supply voltage range is flexible from 6V up to a maximum of 14V. The IC is normally heatsinked to avoid being damaged from overheating.

The actual power output of the IC depends on both the supply voltage and the speaker's impedance. It can be found using the formula shown in the preceding page.

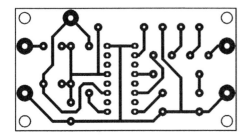
Figure 1.0 Printed Circuit Board

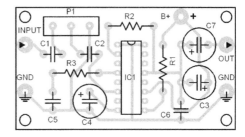
Figure 1.1 Parts Placement

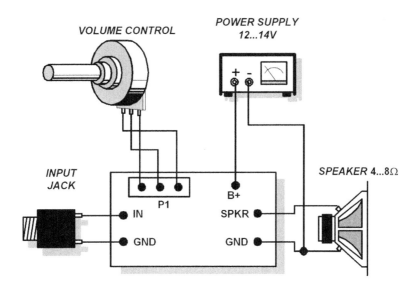
Figure 1.2 External Wirings

Electronic Circuits - 1.0

2 Audio Peakmeter

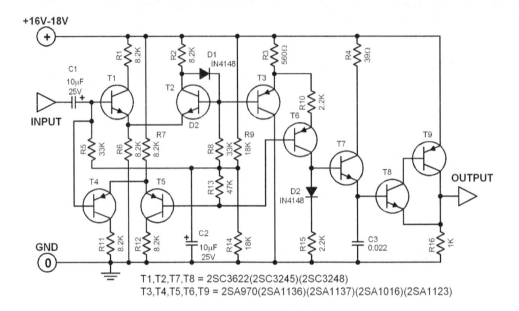

Diagram 2.0 Audio Peakmeter

This circuit can measure the peak level of AC signals regardless of their waveform. It is independent from the signal direction. Both positive and negative going peaks deliver the same result. As you can see in the block diagram, block A is a rectifier for the positive going peaks while block B is a rectifier for the negative going peaks. The rectified peaks are then inverted in block C and added to the rectified signal coming from block A.

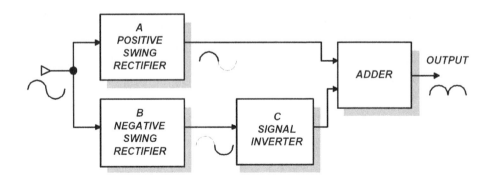

Diagram 2.1 Block Diagram

Audio & Music

A voltage of around 600 mV appears at the circuit's point C. The highest peak value appears at the resistor R16 of the output terminal. A meter can be connected at this point to display the measured value.

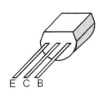

Bottom view

2SC3622
2SC3245
2SC3248
2SA970
2SA1136
2SA1137
2SA1016
2SA1123

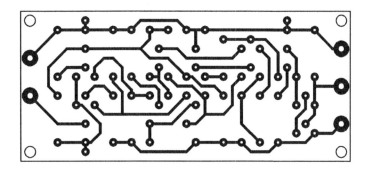

Figure 2.0 Printed Circuit Board Layout

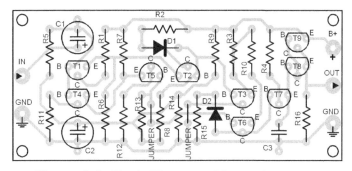

Figure 2.1 Parts Placement Layout

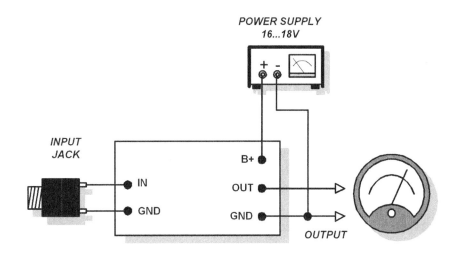

Figure 2.2 External Wirings

Page 15

3 Very Low Noise Mic Amp

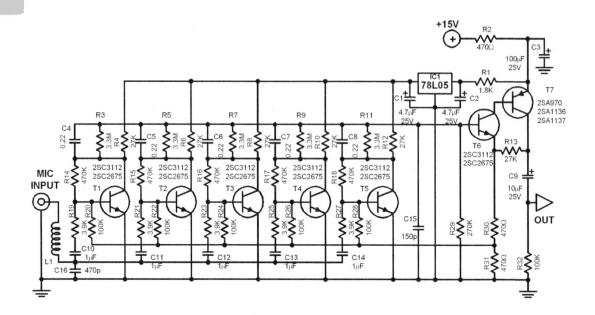

Diagram 3.0 Very Low Noise Mic Amp

This microphone circuit is originally designed for recording sounds from very far distances using microphones with parabolic reflectors. Its noise level is around 7 dB and has a distortion factor of around 1%. This value is less than the distortion factor of a tape recorder. A 130µV input signal gives an output level of 60mV. This figure shows that the microphone amplifier is very sensitive. The amplification factor is around 475. The circuit can also accept input levels up to 8mV. Its bandwidth is from 20 Hz to 45 kHz.

Coil L1 is several turns of magnet wire in a ferrite core. This mic amplifier is about 12 dB better than the mic amp of a good tape recorder. Of course, this circuit can also be used in standard music recording. The quality of the recorded music is much better than the one recorded with an ordinary amplifier.

The resistors used in this circuit are of metal film type to obtain the best results.

Audio & Music

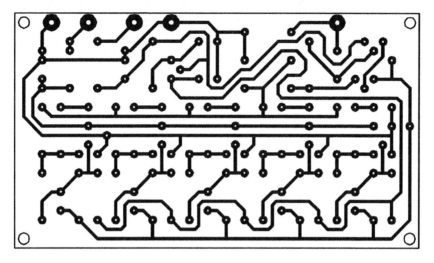

Figure 3.0   Printed Circuit Board Layout

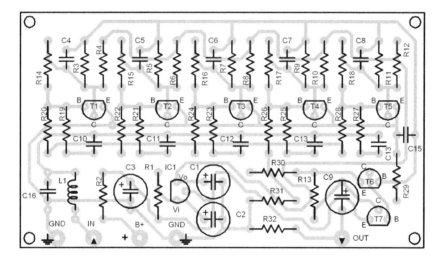

Figure 3.1   Parts Placement Layout

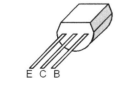

2SA970   2SC3112   78L05
2SA1136   2SC2675
2SA1137

Page 17

4 Signal Clip Indicator

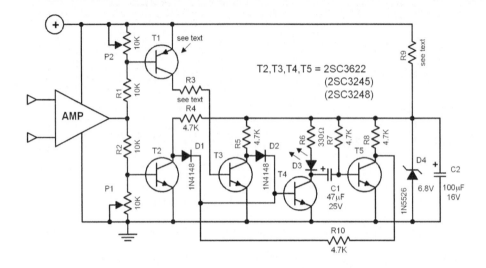

Diagram 4.0   Signal Clip Indicator

This circuit shows the clipping of an output signal coming out of a preamp or a final amplifier through a short lighting of a LED. The supply voltage of the circuit is not critical - it can be either symmetrical or unsymmetrical. This circuit is normally connected as part of the amplifier circuit and uses, therefore, the existing supply lines of the amplifier. The monitored output signal is sampled at the point before the electrolytic output coupling capacitor as shown in the diagram 4.1. The LED lights up at clipping levels in both positive and negative swings of the signal.

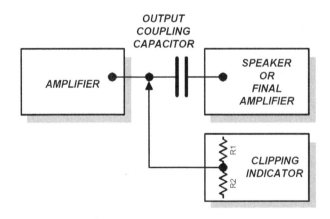

Diagram 4.1   Block Diagram

Audio & Music

The transistor T1 must be selected according to the supply voltage level. If the supply voltage is lower than 40 Volts, use one of the ff: transistor types: 2SA970, 2SA1136, 2SA1137. If the supply voltage is between 40 and 65 Volts, use either 2SC1285 or 2SC1285A. Select the value of R4 to let a current of around 12 mA flow through it.

It is also important to maintain a constant threshold level for the monoflop. To do this, select the the value of resistor R9 so that the current flowing through the zener diode D4 is between 20...24 mA

The calibration of the circuit can be done quickly and accurately by using an oscilloscope. To calibrate: First, connect the oscilloscope at the junction of resistors R1 and R2. Then, inject a positive signal strong enough to cause clipping of the positive peak and adjust P2 until the LED lights up. Conversely, inject a negative signal and adjust P1 until the LED lights up.

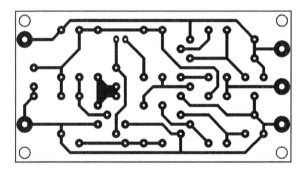

Figure 4.0   Printed Circuit Board Layout

E C B

2SC3622   2SA970
2SC3245   2SA1136
2SC3248   2SA1137

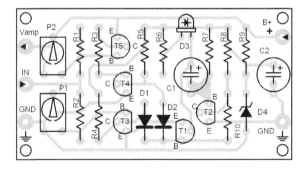

Figure 4.1   Parts Placement

C B E

2SC1285
2SC1285A

Electronic Circuits - 1.0

5   Dyna Audio Compressor

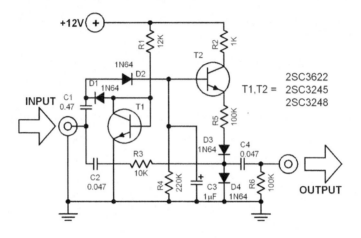

Diagram 5.0   Dyna Audio Compressor

In many audio applications, regulating the audio signal to a constant level is very important. This is true specially for voice modulated radio transmissions. Voice regulation is also often applied in intercom devices or telephone systems. Such a voice level regulator need not be very expensive.

The circuit featured here combines simple design with inexpensive materials. This simple audio compressor uses a straightforward regulation technique to constantly control the amplitude of the output signal. In contrast to common feedback techniques where the output is used to regulate the input signal, the output signal of this circuit is being regulated by the input signal.

Audio & Music

This technique simplifies the overall design of the circuit. Surprisingly, the transistor T2 works as the only active component in the circuit.

This compressor functions very well in intercom systems or in radio transceivers. It can also be used in PA systems or telephone devices such as automatic answering machine.

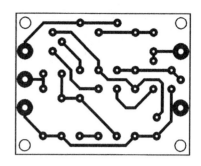

Figure 5.0 Printed Circuit Layout

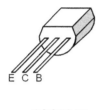

2SC3245
2SC3248
2SC3622

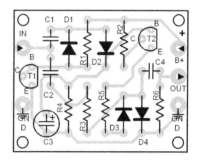

Figure 5.1 Parts Placement

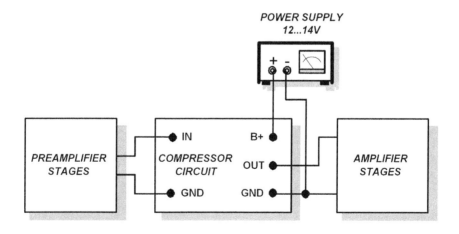

Figure 5.2 Installation Wirings

Electronic Circuits - 1.0

6 Stereo Audio Mixer

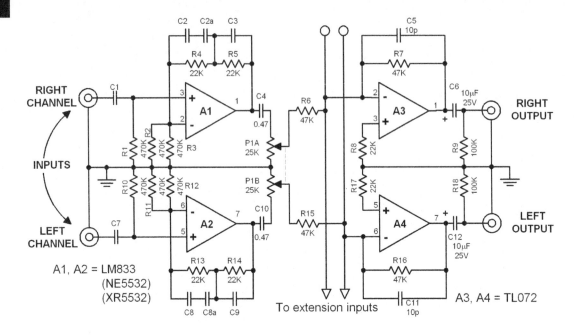

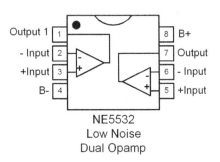

NE5532
Low Noise
Dual Opamp

Diagram 6.0   Stereo Audio Mixer

This is a mixer circuit which can mix stereo sources. The component values for the input circuit varies according to the type of instrument connected to it. Table 6.0 shows the necessary component values for every input instrument. You will notice that if you design the input for a tape instrument you don't need the A1/A2 IC and the discrete components around it. You simply connect the source line to the capacitor C4 or C10.

Table 6.0

input device	C1/C7	C2/C8	C2a/C8a	C3/C9	R1/R10	R2/R11	R3/R12	R4/R13	R5/R14
Mic-low ohm	10mF/10.2V	-	-	10p	680W	21K	-	short	100K
Mic - Hi ohm	0.47	-	-	10p	22K	1K	-	short	100K
Tape	remove A1 & A2, connect directly to C5 or C10								
Phono	0.22	0.0015	0.0015	0.0033	47K	2.2K	2.2K	100K	1M

Audio & Music

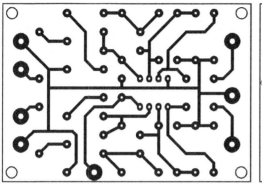

Figure 6.1 Printed Circuit Layout of the Input Preamp Module

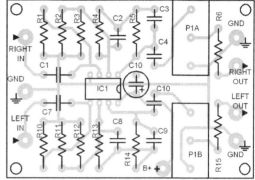

Figure 6.2 Parts Placement Layout of the Input Preamp Module

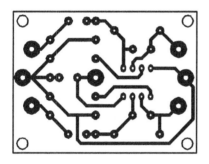

Figure 6.3 Printed Circuit Layout of the Main Mixer Module

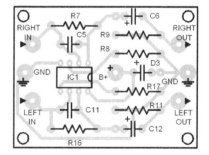

Figure 6.4 Parts Placement Layout of the Main Mixer Module

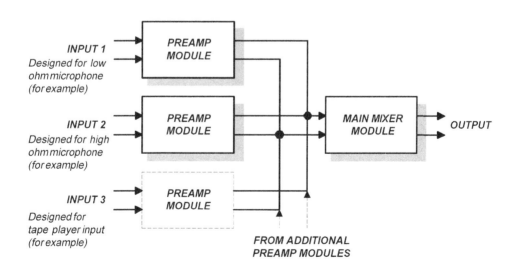

Diagram 6.1   Installation Diagram

Page 23

7 Tunable Filter Circuit

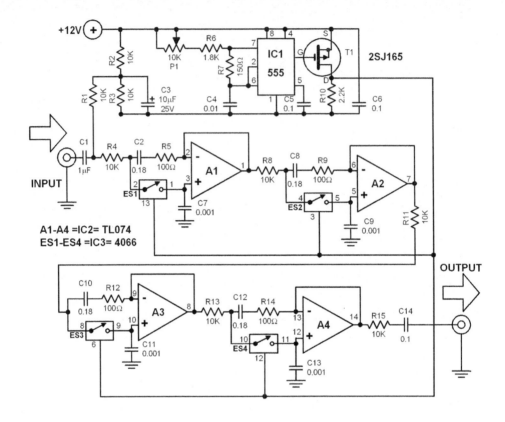

Diagram 7.0 Tunable Filter Circuit

This circuit uses a clock generator to control four analog CMOS switches. The switches in turn vary the input resistance of each opamp. These analog switches are controlled by a 555 clock generator with a duty cycle that is variable from 1:10 up to 100:1. If the analog switch is closed, the input resitance is around 60 ohms and when it is open, the resistance is almost infinite. If, for example, the switch is clocked with a duty cycle of 0.5, the resulting input resistance is 1/(0.5/60) = 120 ohms. If it is 0.25, the resistance is 240 ohms. It shows that every half of the dutry cycle results to a doubling of the input resistance.

The frequency of the clock signal must be several times higher than the highest audio frequency to be processed to avoid hearing the resulting interference between the clock and the audio signals. The amplification is around 40 and is dependent on the clock frequency.

Audio & Music

2SJ 165
Bottom view

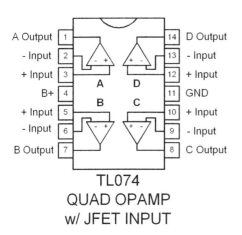

TL074
QUAD OPAMP
w/ JFET INPUT

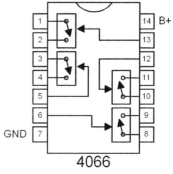

4066
Quad Digital/Analog Switch

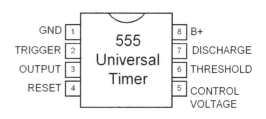

555 Universal Timer

8 Channel Balance Indicator

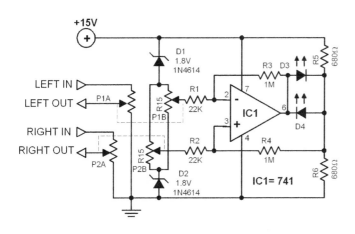

Diagram 8.0   Channel Balance Indicator

Are you tired and frustrated of eternally trying to balance your stereo amplifier? If so, why not let electronics listen to the amplifier's output for you, and tell you whether the two channels are balanced or not?

Such automatic listener and balance analyzer is featured here. This simple circuit provides an optical way of telling whether the two (left and right) channels are balanced or not.

Page 25

Electronic Circuits - 1.0

The potentiometer pair P2a and P2b is mechanically coupled to each other as well as the pair P1a and P1b as shown by the dashed lines in the circuit diagram. There are also paired potentiometers available in electronics supply stores. I recommend buying such ready made pots. They are accurate and will help you minimize your efforts in constructing this circuit.

After building the circuit, wire it properly. Important: turn down the volume of the amplifier before connecting this analyzer circuit. Afterwards, connect the input lines of the balance analyzer to the two output channels of your amplifier. Turn on the amplifier and input a signal to it. Ideally, the input signal must be a clean sine wave with a constant amplitude such as the one produced by a signal generator. Slowly, adjust the volume of one channel until a LED lights up. Adjust the volume of the other channel until the LED turns off.

When the two channels are balanced, the LEDS remain turned off. Otherwise, one of the LEDs will light showing which channel is stronger.

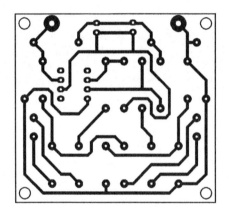
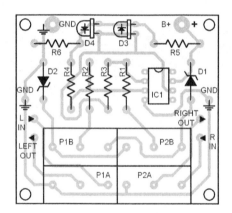

Figure 8.0 Printed Circuit Layout   *Figure 8.1* Parts Placement

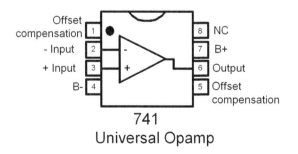

741
Universal Opamp

Audio & Music

9 Tweeter Guardian

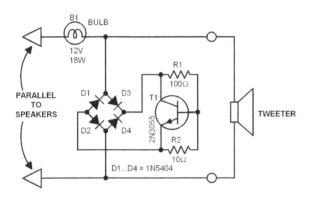

Diagram 9.0 Tweeter Guardian

Small but very useful. This circuit provides protection to your tweeter specially when they are always driven near the maximum level. The first circuit uses a simple bulb as a load. This bulb will glow when the signal going to the tweeter reaches a certain preset threshold level.

The bulb functions like a positive temperature coeficient (PTC) resistor - meaning its resistance increases proportionally to its temperature. At around 5.5, the transistor 2N3055 will conduct and shorts the tweeter's input line to ground, thereby preventing the overloading of the tweeter.

The second circuit is an improved version. A fixed resistor replaces the bulb and the switch transistor is a darlington pair with a delay capacitor. The circuit still functions like the first circuit. However, due to the capacitor, the circuit will not react to a single overload peak.

This circuit is usually installed inside high power boxes like those in disco applications where equipments are commonly driven to their maximum ratings.

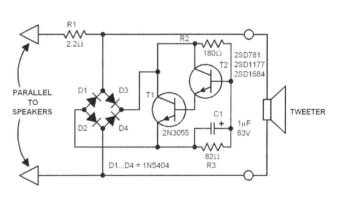

Diagram 9.1 Tweeter Guardian
(improved version)

Page 27

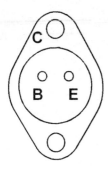

2N3055
Bottom view

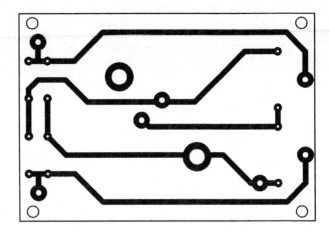

Figure 9.0   Printed Circuit Layout

2SD781
2SD1177
2SD1684

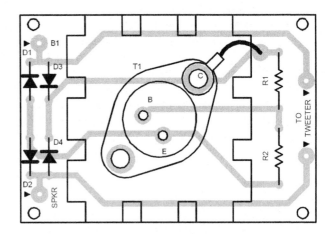

Figure 9.1   Parts Placement Layout

2SA970	2SC3622
2SA1136	2SC3245
2SA1137	2SC3248

Parts List for version 2

T1 = 2N3055
T2 = 2SD781 (2SD1177)(2SD1684)

D1...D4 = 1N5404

R1 = 2.2 Ω
R2 = 180 Ω
R3 = 82 Ω
C1 = 1 µF/63 volts

10 Low Noise Preamp

Audio & Music

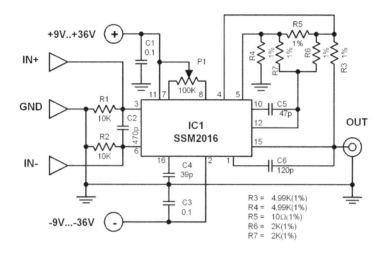

Diagram 10.0 Low Noise Preamp

This is a single IC low noise preamp made possible by using the SSM2016 from PMI. It has a symmetrical input and extremely low noise characteristics. It is highly capable of amplifying signal sources with low internal resistance (less than 1 kiloohms) like moving coil microphones and dynamic microphones with impedances ranging from 10 ohms to 600 ohms.

The amplification is adjustable by changing the resistance of R5 with a value between 3.5 and 1000 ohms. The amplification A can be found with the ff: formula:

$$A = (R3+R4)/R5 + (R3+R4)/(R6+R79)$$

It can be simplified to A= 10K/R+3.5 using the given component values. If for example R5 = 10 ohms, then the amplification is 1000. The offset voltage at the inputs can be set to zero through potentiomer P1. It is highly recommended to use metal film resistors in constructing the circuit.

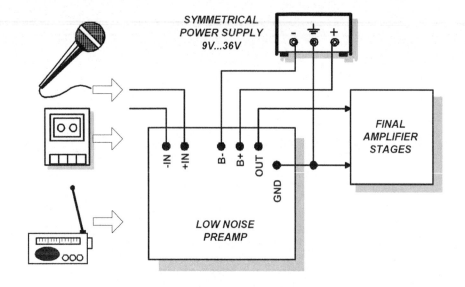

Figure 10.0   External Wirings

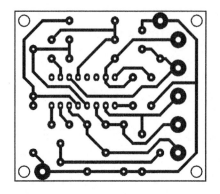

Figure 10.1   Printed Circuit Layout

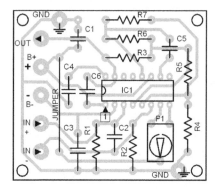

Figure 10.2   Parts Placement Layout

Parts List

IC1 = SSM2016

C1,C3 = 0.1 uF
C2 = 470 pF
C4 = 39 pF
C5 = 47pF
C6 = 120pF

R1,R2 = 10K Ω
R3,R4 = 4.99KΩ 1%
R5 = 10Ω 1%
R6,R7 = 2kΩ 1%

11 Envelope Sampler

Audio & Music

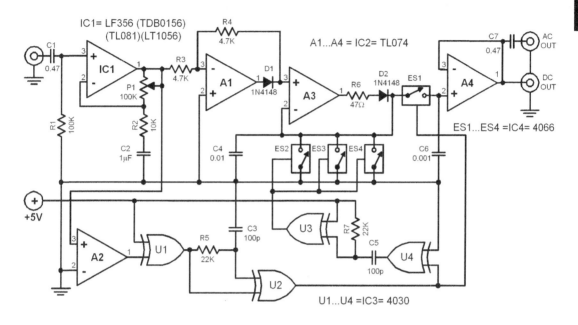

Diagram 11.0 Envelope Sampler

This is an AM demodulator which is not affected by phase errors in the demodulated signal. Phase error is a type of error commonly found in simple diode/low pass filter demodulators. At the input stage is an AC amplifier which is adjustable through P1. A1 and A3 are the rectifiers that charges C6 with the maximum signal voltage. The analog switches are controlled by a clock signal. The clock signal is derived from the carrier frequency. In this way the sampler can be applied in different devices like facsimile, radio and speed processors since its clock frequency is directly controlled by the received carrier frequency.

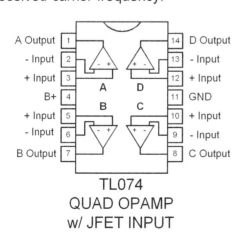

TL074
QUAD OPAMP
w/ JFET INPUT

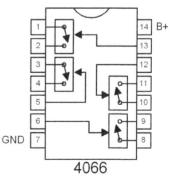

4066
Quad Digital/Analog Switch

12 Microphone Processor

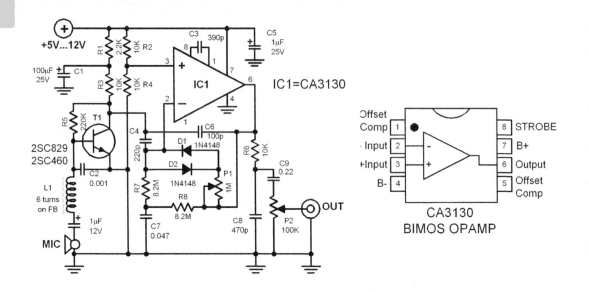

Diagram 12.0 Microphone Processor

Audio processors are usually used in paging systems, in wireless intercom and the likes to amplify the microphone signal to a certain level. This can be done by using either a compressor or a limiter circuit. Although a compressor has lower distortion characteristics, it is a very complicated circuit. The limiter is simpler to construct, but it has a relatively high distortion level. Intermodulation distortion is high in a limiter circuit and in order to effectively use a limiter, you have to suppress the intermodulation interference as much as possible. This can be done by automatically changing the limit frequency according to the strength of signal input. The circuit featured here does just that.

This circuit has an amplifier with a very high input impedance. When the input signal is still low, the diodes do not conduct yet. In this situation the limit frequency is still dependent on R1 and C1. Once the diodes conduct, (it happens when the input signal has increased), the input impedance of the amplifier decreases thereby shifting the limit fequency to a higher value. The lower frequencies are then amplified less thus making the audio signal more understandable. The signal processed this way is much better than the one which is just simply "clipped". This circuit is also applicable to process music signals.

The values of C6, C7 and C8 are given in the following table according to the application of the circuit.

	C6	C7	C8
music applications	-	47n	470p
voice applications	100-220p	0 - 4n7	4n7

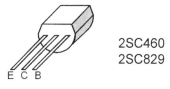

2SC460
2SC829

13 HIFI Audio Mixer Module

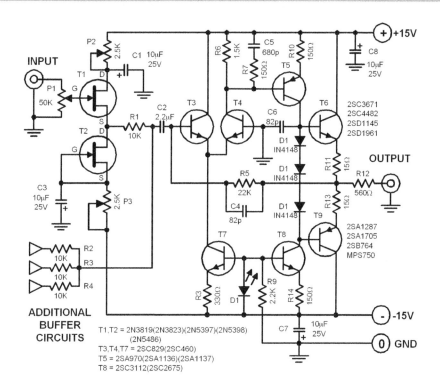

Diagram 13.0 HIFI Audio Mixer Module

High fidelity mixers deliver high dynamic and low noise characteristics but expensive. Conventional mixers on the other hand are constructed with cheap OPAMPs which are quite noisy. The noise problem can be avoided by buffering the inputs and instead of using OPAMPs, discrete components are used for the amplifier stages.

Electronic Circuits - 1.0

These design considerations are applied in the circuit featured here. Transistors T1 and T2 are the buffers. The input impedance is dependent on the setting of P1. Transistors T3 up to T8 build the amplifier stage and are HF type transistors. HF transistors have lower noise factor than their AF counterparts.

The described circuit is designed not only for mono inputs. It can also be expanded to process stereo inputs. To do this, just construct two identical circuits and use them as separate modules for both left and right inputs. One thing to be very careful though in constructing the stereo modules, make sure that the wirings are not mistakenly crossed between the two identical circuits.

Technical characteristics:

Frequency range = (-3 dB) 10 Hz ... 80 kHz
Signal to noise ratio (9V output/20 kHz bandwidth) = 100 dB with 10 buffer stages
Maximum output signal = 12 Vpp

The buffer circuit must be duplicated for every additional input channel and connected to C1 in the amplifier stage.

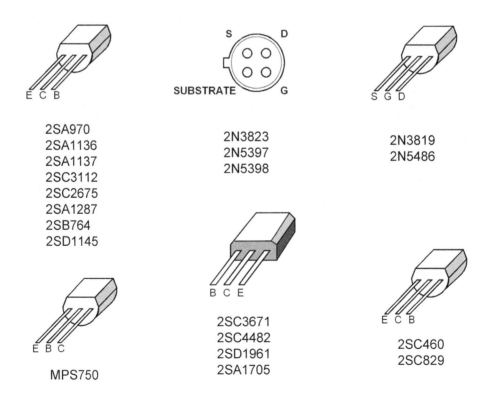

HOBBY & SHOP

36 Digital Bandpass Filter
38 Transistor Solar Cells
39 Hybrid Cascaded Transistor
40 Adjustable Dummy Load
42 Optocoupler
43 Automatic Battery Charger
44 Voltage/Frequency Converter
46 Music in a Chip
48 Ultrasound Receiver

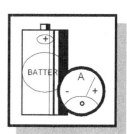

Electronic Circuits - 1.0

14 Digital Bandpass Filter

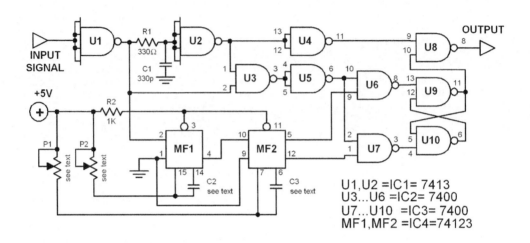

Diagram 14.0 Digital Bandpass Filter

This filter lets digital frequencies within its passband to pass though with the help of two monoflops. A bandpass filter has two definite cut-off frequencies: fmax and fmin. The bandwidth b is fmax-fmin while the center frequency is fmin+b/2 or fmax-b/2.

The upper cut-off frequency fmax is determined by the pulse length (t1) of the monoflop M1.

$$t1 = Tmin \frac{b}{fmax}$$

Parts List:

Resistors:
R1 = 330Ω
R2 = 1K
P1,P2= 5...50K
see text

Capacitors(ceramic):
C1 = 330p/50V
C2,C3 = see text

IC:
U1,U2=IC1 = 7413
U3...U6=IC2 = 7400
U7...U10=IC3 = 7400
MF1,MF2=IC4 = 74123

Hobby & Shop

The lower cut-off frequency (fmin) comes from the sum of the pulse length of the monoflop 1 and monoflop 2.

$$f_{min} = \frac{1}{t1+t2}$$

The pulse length (t) is determined by RC components P2/C2 and P1/C3 and can be found mathematically by:

$$t = 0.32 * C * (P+700)$$

The value of P can be chosen from 5 ohm up to 50K.

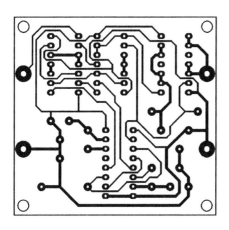

Figure 14.0 Printed Circuit Layout

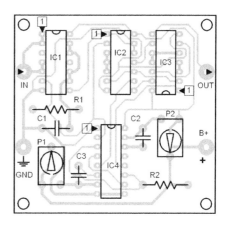

Figure 14.1 Parts Placement Layout

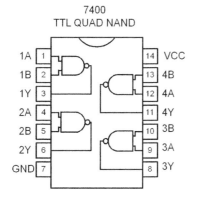

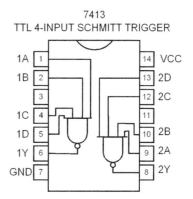

Electronic Circuits - 1.0

15 Transistor Solar Cells

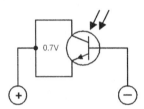

Diagram 15.0   Transistor Solar Cells

Almost every electronic hobbyist has a pile of old or defective 2N3055's in his hobby schack. These transistors can be recycled to provide energy - solar energy. A power transistor like 2N3055 can be used as a solar cell by cutting or sawing off its metal top, exposing its silicon chip, and then placing it under the sun. Even a defective transistor can be used for this purpose as long as two of its terminals still create a diode junction.

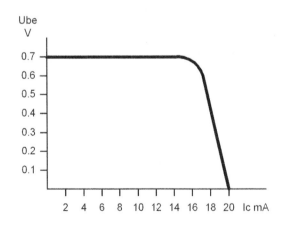

Diagram 15.1   Output voltage to load current of the 2N3055 used as a solar cell.

A transistor used as a solar cell produces voltages up to 0.7V. However, this output voltage is dependent on the load current. The graph on the left shows the relation of its output voltage to the load current. This current output can be increased by using a non-defective transistor, and connecting its collector and emitter together in parallel as shown in the diagram 15.1. To increase the voltage level, you can connect several such transistors in series. By joining them in parallel, you can increase the current output.

16 Hybrid Cascaded Transistor

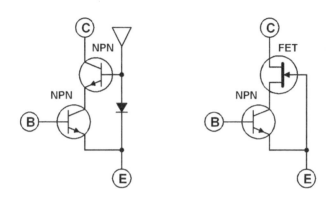

Diagram 16.0 Hybrid Cascaded Transistor

Two or more transistors can be wired together to create a new single transistor. This hybrid transistor has better characteristics compared to each one of the transistors used. The advantages are higher collector impedance and lower sensitivity to noise at the input. The cascaded transistors approximate the function of a constant current source. The first circuit shown uses two transistors, and the second one uses a combination of transistor and FET. The FET combination offers a much less critical DC voltage stabilization.

Typical use of the circuit is in applications where a constant source of current is important, like for example constant current battery chargers or reference current modules.

2SC3622　2SA970
2SC3245　2SA1136
2SC3248　2SA1137

17 Adjustable Dummy Load

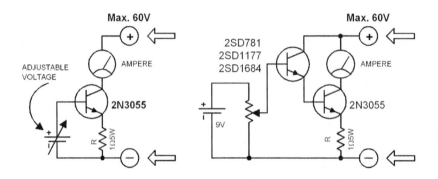

Diagram 17.0 Adjustable Dummy Load

This dummy load can be very helpful in testing power supplies, storage batteries, and similar devices. It has a variable load resistance and enough current handling capacity. Ordinary dummy loads are usually made of fixed value resistors. They are cheap, readily available, and easy to assemble. But for power dissipations above 10 watts, it is quite difficult to find the appropriate resistors. Their load resistance value is also fixed.

One excellent solution to this problem is using a dynamic dummy load. This dummy load not only can dissipate high power levels, but its load resistance can also be varied. The circuit featured here uses a 2N3055 transistor with an emitter resistance of 1 ohm per 5 watts. It functions as a variable current sink. The current flowing through the transistor can be varied by applying a voltage at its base. The load current value (I_L) can be found with the ff: formula:

$$I_L = \frac{(U_V - U_{BE})}{R}$$

Where U_V = base control voltage
U_{BE} = base-emitter voltage
R = resistance at the emitter

The 2N3055 transistor must be heatsinked. The heatsink helps the transistor handle loads up to 45 watts.

Hobby & Shop

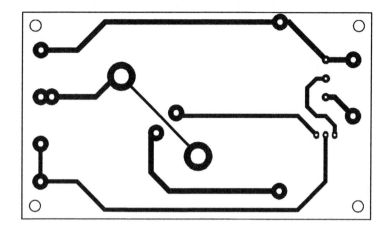

Figure 17.0 Printed Circuit Layout

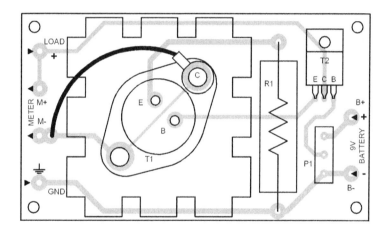

Figure 17.1 Parts Placement Layout

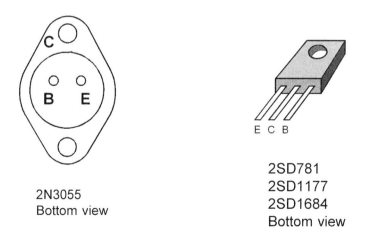

2N3055
Bottom view

2SD781
2SD1177
2SD1684
Bottom view

Electronic Circuits - 1.0

18 Optocoupler

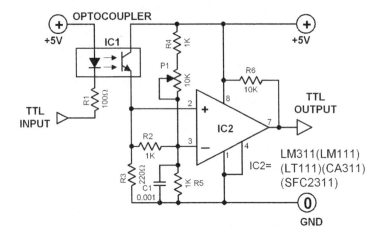

Diagram 18.0   Optocoupler

When a computer must control an external device -specially when the device is powered by AC voltage - the computer must be electrically isolated from the external device to prevent possible damage to the computer, when something goes wrong with the external device. The simplest way to do this is to use a transformer. However, by relaying short control pulses or high freguency signals, an optocoupler works best.

The circuit featured here uses a comparator following the phototransistor. The comparator has a TTL compatible output. Potentiometer P1 varies the threshold level of the comparator.

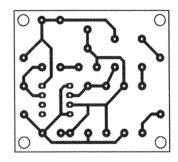

Figure 18.0
Printed Circuit Layout

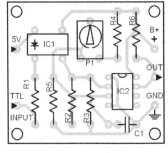

Figure 18.1
Parts Placement Layout

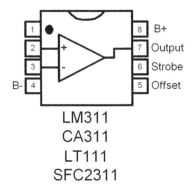

LM311
CA311
LT111
SFC2311

19 Automatic Battery Charger

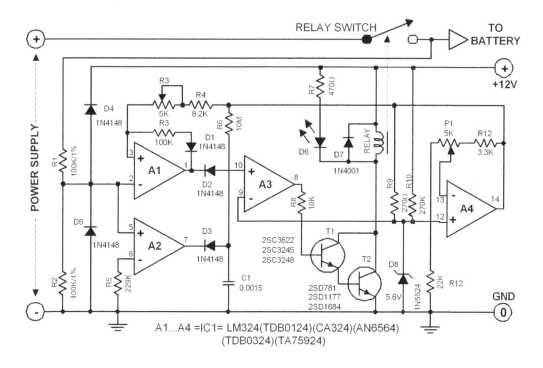

Diagram 19.0 Automatic Battery Charger

The charging technique applied in this circuit is very simple and straightforward. The circuit closes a relay to connect the charging line to the battery. Once the battery voltage reaches 13.8V, the relay is deactivated and the charging is stopped. The relay will be closed again (charging resumed) when the battery voltage is less than 12.6V. The relay must be able to handle at least 5 amperes.

To calibrate the circuit, first set P1 to minimum so that the relay closes and the battery is charged. Once the battery voltage is 13.8 volts, adjust P1 until the relay opens. Use the battery normally to discharge it until the voltage is around 12.5V, then set P2 so that the relay closes once again.

2SD781
2SD1177
2SD1684
Bottom view

20 Voltage/Frequency Converter

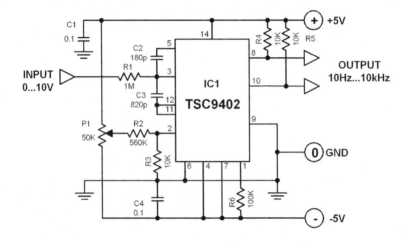

Diagram 20.0   Voltage/Frequency Converter

A voltage to frequency converter with a control range of 1:1000 can be easily constructed by using the IC TSC9402. The given component values in the circuit produces a conversion factor of 1kHz/1V. The input voltages from 10 mV up to 10 V are converted to frequencies 10 Hz up to 10 kHz. This conversion can be changed if desired through the potentiometer P1.

The circuit has two outputs: a sharp pulse comes out from pin 8 with the main output frequency. A clean squarewave comes out of pin 10 but its frequency is half of the output frequency at pin 8. **Calibration**: use an accurate frequency counter and adjust P1 so that an input level of 10 mV will produce a signal of 10 Hz at the output.

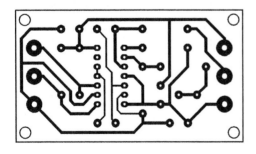   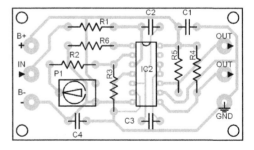

Figure 20.3   Printed Circuit Layout   **Figure 20.4**   Parts Placement

Hobby & Shop

One area of application for this circuit is the telemetry. In this case, telemetry of voltage values. By converting the voltage value to a frequency, the value can be transmitted either via a coaxial cable or radio. Wireless telemetry method using radio, can be realized by using the output of this circuit to modulate the carrier signal of a transmitter. It is not possible to use directly the output frequency of this circuit since it produces only a maximum of 10 kHz (equivalent to a measured voltage of 10 volts at its input).

The modulation technique can be any of the available industry standard modulations. It can be AM, FM, SSB, FSK, etc.

The following diagram shows how to use the circuit for remote measurements using a coaxial cable. The technique is limited to several meters though. The distance can be increased by using a signal amplifier.

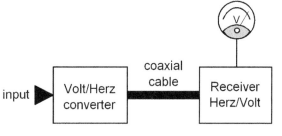

Figure 20.1   Telemetry via coaxial cable

The diagram on the right shows the circuit modulating a transmitter to transport the measured value via radio signals. This technique allows remote telemetry over very long distances. If the signal is repeated via a satellite transponder, the telemetry can be done worldwide.

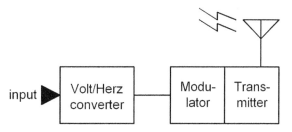

Figure 20.2   Wireles telemetry

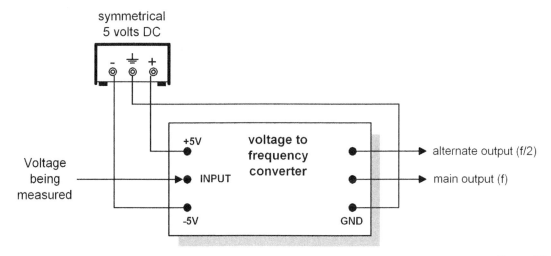

21 Music in a chip

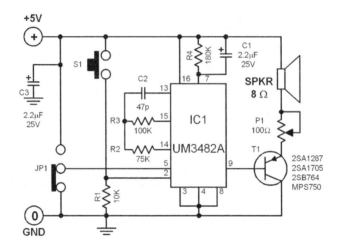

Diagram 21.0 Music in a chip

This 1-chip module can generate 9 different melodies. It is normally used as a doorbell, an acoustic signalling device or as a telephone "wait" melody.

Once switch S1 is pressed the IC plays the melodies one after the other. It will stop after the last melody when the JP1 terminal is connected to ground by a jumper. Alternatively, when JP1 is connected to the +V, the IC will play continously as long as the switch S1 remains closed. Combining several pins produces different effects as shown in Table 21.0. A single battery can be used to power this module since its current consumption is negligible.

Hobby & Shop

Table 21.0				
Pin 2	Pin 3	Pin 4	Pin 5	Effect
0	x	x	x	Standby
1	0	0	0	first-last-stop
+ Pulse	0	0	1	first-last-repeat first
+ Pulse	1	0	0	current-stop
1	1	0	1	repeat current
1	0	+ Pulse	0	next-last-stop
1	0	+ Pulse	1	next-last-repeat first
1	1	+ Pulse	0	next-stop
1	1	+ Pulse	1	next-repeat same

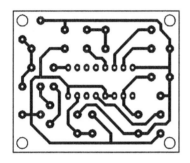 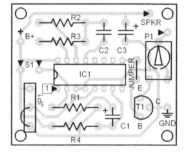

Figure 21.0 Printed Circuit Layout   *Figure 21.1* Parts Placement

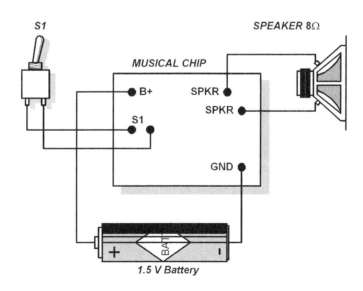

Figure 21.2 External Wirings

22 Ultrasound Receiver

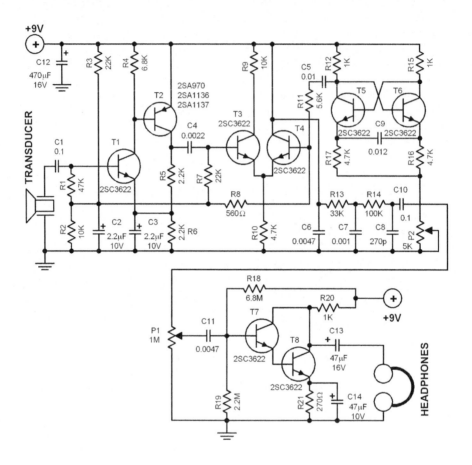

Diagram 22.0 Ultrasound Receiver

The audio frequencies between 15 kHz and 18 kHz is normally not audible to humans. This audio range is also called ultrasound. In order to hear it, it must first be converted to a frequency range which is audible to humans.

The circuit featured here mixes the ultrasound with a frequency generated by a BFO circuit. The result is a mixture of several frequencies from which one (the difference) can be clearly heard. The undesirable signals are filtered out by a 4 kHz low pass filter. The transducer is a special ultrasound sensor. Portable batteries can be used to power this circuit.

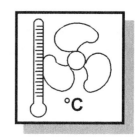

HOUSE & CAR

50 Atmospheric Disturbance Detector
52 Electronic Fuse
54 Lamp Dimmer/Speed Regulator
56 Intercom
57 Optically Coded Key
59 Thermometer
60 Automatic Lamp Regulator
62 Universal Timer Circuit
64 Smoke Alarm
65 Electronic Selector Switch
66 Infra-Red Switch
69 DC Flourescent Lamp
71 Thermonitor
73 Live Wire Sensor

23 Atmospheric Disturbance Detector

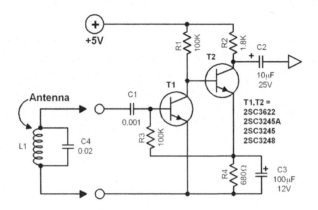

Diagram 23.0   Atmospheric Disturbance Detector

Atmospheric disturbances, air disturbances, typhoons, and in-fact atomic bomb explosions produce not only massive air movements but also magnetic oscillations. These magnetic oscillations can be detected from long distances with the help of a sensitive amplifier and a special antenna. Some people dubbed this circuit as "storm detector".

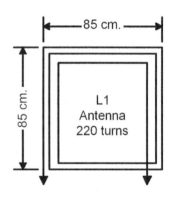

Loop Antenna L1

The detector circuit is actually very simple. The most important part is the antenna (shown in the diagram as L1). It is made of 220 turns magnetic wire wound side by side around a square wooden frame. The square frame's dimension is 85cm by 85 cm.

One 0.022 µF capacitor (labeled as C4 in the diagram) must be soldered in parallel to the ends of the coil winding. This coil/capacitor combination produces a resonance of around 4 to 5 kHz. The output of the detector can be connected to either an oscilloscope or a clocked oscillograph. This circuit is most often used as a typhoon detector.

House & Car

2SC3622
2SC3245
2SC3248

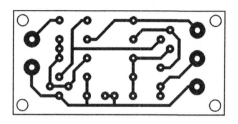

Figure 23.0 Printed Circuit Layout

2SA606   2SC696
2N2303   2SD1639
2N1990

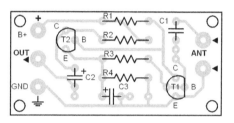

Figure 23.1 Parts Placement Layout

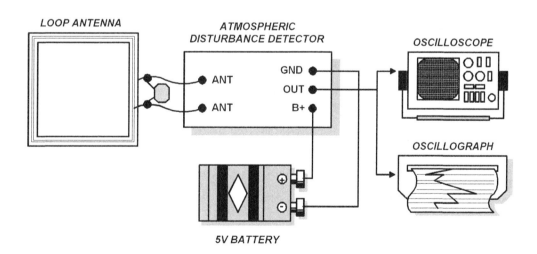

Figure 23.2 External wirings

Electronic Circuits - 1.0

24 Electronic Fuse

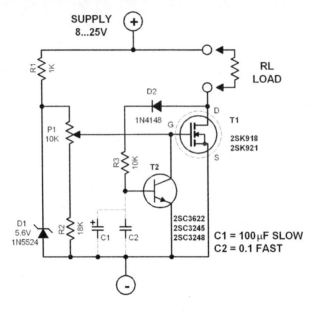

Diagram 24.0   Electronic Fuse

Tired of replacing fuses everytime they blow? Use an electronic fuse. It will still "blow" and protect your gadgets, but you don't need to buy a replacement fuse. This "fuse" can be "rehabilitated" by just resetting it. This circuit is the electronic equivalent of a circuit breaker. Though, you might say a circuit breaker is cheaper and easier to install, constructing this one will enhance your understanding of the sensor technique.

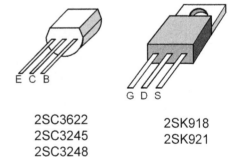

2SC3622
2SC3245
2SC3248

2SK918
2SK921

This electronic fuse uses an N-channel power FET as the high current sensor. The FET functions at the same time as the actual circuit breaker that cuts-off the ground line of the load, once the current exceeds the maximum allowed value.

WARNING!
The maximum current capacity of this "fuse" is 5 amperes.

The cut-off current can be adjusted through the pot P1 from 0 to 5A. Remember, the circuit can only handle a maximum current load of 5 amperes. Do not overload this circuit if you do not want to fry it. The power FET becomes hot in time, so it must be heatsinked.

Use also the proper wire gauges in wiring the circuit to its inputs and outputs. Using the wrong wire gauge will damage the circuit. A relatively extensive test is needed before you release this "fuse" for a real long time protection job.

You must connect either C1 or C2 to the base of transistor T2. C1 will cause the fuse to blow "softly" while C2 will cause it to blow instantly. Use different resistor values (high wattage!) in series with an ampere meter to simulate different current loads. Adjust the pot P1 until the fuse "blows".

Resetting the fuse is simple: Just disconnect its power supply, and the circuit is reset and ready to protect your gadgets again.

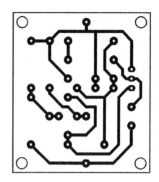

Figure 24.0 Printed Circuit Layout

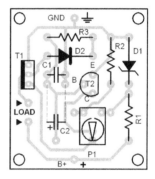

Figure 24.1 Parts Placement

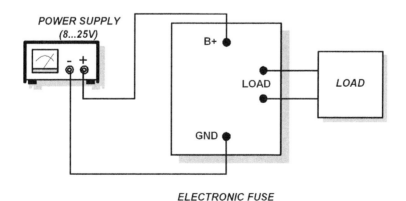

Figure 24.2 External Wiring

24 Lamp Dimmer/Speed Regulator

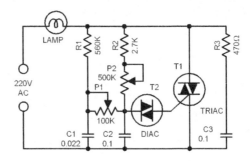

Diagram 25.0 Lamp Dimmer/Speed Regulator

This is not just another dimmer circuit. The improved design of this dimmer circuit enables its triac to be triggered with lower voltage values. Normally, the triggering voltage of a triac is around 70V. In some applications though, it is desirable to use a lower voltage in triggering the triac. A trick is used to achieve this effect in this circuit: the addition of the components R1, C1 and P1. Due to this, the trigger value can be pulled down to around 35V.

Construct the circuit and put it in a highly insulating box. Remember, the circuit is working with 220 volts! For P2 use a special potentiometer with a plastic shaft.

After constructing the circuit, check all wirings and isolations again before you plug it in the power line.

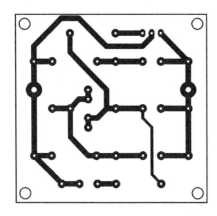

Figure 25.0 Printed Circuit Layout

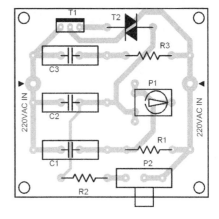

Figure 25.1 Parts Placement

Turn the pot P2 first to its leftmost position. Then plug the load (bulb, electric drill, motor, etc.) before you plug the dimmer into the power line.

Pot P1 must be set in such a way that the lamp or the drill machine will almost start even though speed control P2 is at its leftmost position.

There are lots of practical applications for this circuit. One typical application is controlling the speed of an ordinary electric drill machine. It can also be used to control the speed of a sewing machine, or the electric motor in a model boat, etc.

Parts List:

R1 = 560K
R2 = 2.7K
R3 = 470Ω
P1 = 100K trimmer pot
P2 = 500K pot (plastic shaft)

C1 = 0.022/600V ceramic
C2,C3 = 0.1/600V ceramic

T1 = Triac 400V/3A
T2 = Diac

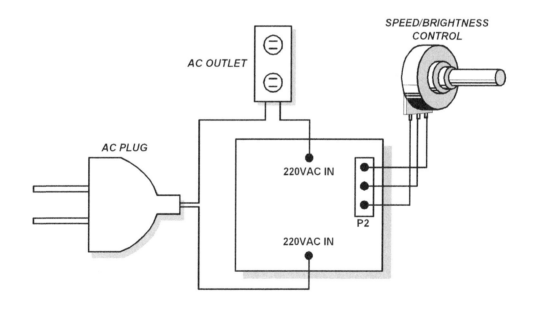

Figure 25.2 External Wirings

**DANGER OF ELECTROCUTION!
THE CIRCUIT IS CONNECTED TO 220 VAC LINE!
USE A POTENTIOMETER WITH PLASTIC SHAFT FOR P2:**

Electronic Circuits - 1.0

26 Intercom

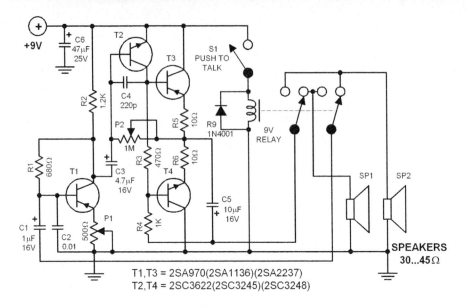

Diagram 26.0 Intercom

This intercom uses an ordinary speaker to function as its microphone. Once the button is pressed, the relay toggles, and the speakers interchange their functions. One works as a normal speaker and the other one works as microphone.

The circuit is composed of a complementary final amplifier (T3 & T4) and two voltage preamplifiers (T1 & T2). Pot P1 controls the volume.

The voltage at the common junction of resistors R5, R6 & P2 must be set to one half of the supply voltage.

A lower impedance speaker with a series resistor can replace the suggested type, but will deliver lower sound quality.

Parts List:

R1 = 680Ω
R2 = 1.2K
R3 = 470Ω
R4 = 1K
R5 = 10Ω
P1 = 500 Ω

C1 = 1µF/16V
C2 = 0.01/50V ceramic
C3 = 4.7µF/16V
C4 = 220p/50V ceramic
C5 = 10µF/16V
C6 = 47µF/25V

T1,T3 = 2SA970(2SA1136)(2SA1137)
T2,T4 = 2SC3622(2SC3245)(2SC3248)
Relay (DPDT) 9V
Speakers (30...45W) = 2 pcs.
Switch (SPST) = 1 pc.

2SA970   2SC3622
2SA1136 2SC3245
2SA1137 2SC3248

27 Optically Coded Key

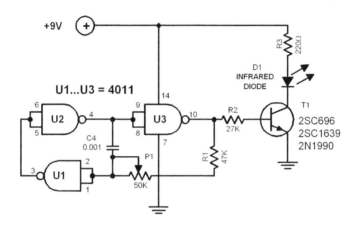

Diagram 27.0 Optically Coded Key (Transmitter Module)

The optical key works as an infrared door opener for house, garage and the likes. Diagram 27.0 functions as the key. It is a transmitter module which sends out a modulated infrared (IR) beam. The modulating frequency can be adjusted through P1.

Diagram 27.1 is the actual lock (receiver module). It uses a phototransistor to receive the IR beam coming from the key transmitter module. It compares the modulation frequency of the transmitter beam to its internal frequency. When the two frequencies match, the relay is activated, thereby opening whatever elecric door lock is connected to it.

The circuit can be modified to switch a car alarm on or off by replacing the relay with a toggle flip-flop circuit.

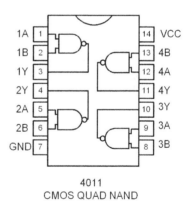

4011
CMOS QUAD NAND

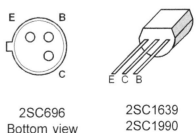

2SC696
Bottom view

2SC1639
2SC1990

Electronic Circuits - 1.0

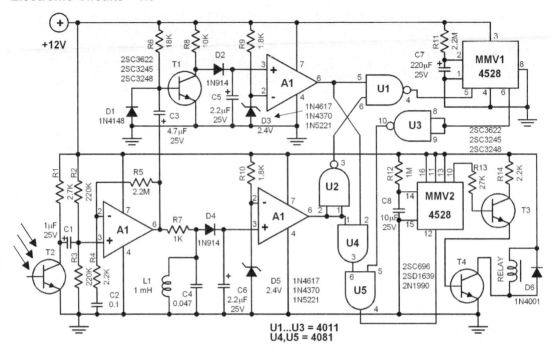

Diagram 27.1 Optically Coded Key (Receiver Module)

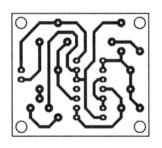

Figure 27.0
Printed Circuit Layout

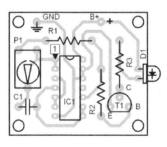

Figure 27.1
Parts Placement

2SC696   2SC3622
2SC1639 2SC3245
2SC3248

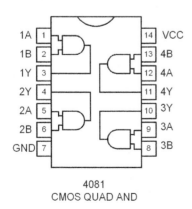

4081
CMOS QUAD AND

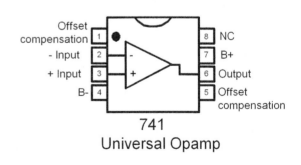

741
Universal Opamp

28 Thermometer

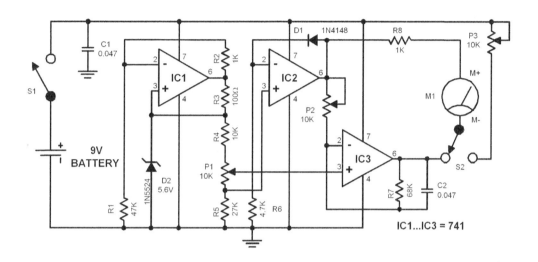

Diagram 28.0 Thermometer

Most diodes exhibit temperature dependency. This characteristic can be exploited to enable temperature measurements through electronic means. In the circuit featured here, an ordinary diode (labeled as D1 in the diagram) is used as the temperature sensor. When a constant forward current flows this diode, the voltage in the diode is proportional to its temperature. This means that in order for the electronic thermometer to function correctly, a constant current must be available. This constant current can be taken from a constant reference voltage source. A very stable reference voltage is provided in the circuit by wiring an IC (labeled as IC1 in the diagram) with an ordinary zener diode(labeled as D2 in the diagram). This combination works as a super-zener diode. A 5.6V zener diode is selected because this type is the least temperature dependent.

The temperature change at the sensor diode results to an output at the IC2 of around -2mV/°C. This voltage output is then amplified by IC3 and delivered to the meter.

Calibration: Potentiometer P1 sets the lowest temperature to be measured. Potentiometer P2 sets the full deflection at the highest temperature to be measured.

29 Automatic Lamp Regulator

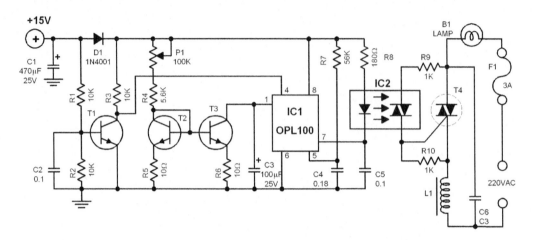

Diagram 29.0 Auto Lamp Regulator

This circuit automatically adjusts the brightness of a lamp according to the intensity of the surrounding light. A sensor measures the total surrounding brightness and the lamp's light intensity is controlled in the opposite direction. When the surrounding light is much brighter than usual (as regulated through the pot P1) the circuit dims the lamp. On the other hand, when the surrounding light is darker than usual, the circuit increases the current supply to the lamp increasing its light intensity. This negative feedback regulation technique ensures a relatively constant overall room illumination.

The sensor used in this circuit is an OPL100. It has a photodiode and an integrated regulator. The regulation control is passed on to the actual lamp circuit through an optocoupler that provides the necessary electric and galvanic isolation. This optocoupler technique is very important to increase the safety level of the circuit. Potentiometer P1 provides manual brightness adjustment.

Follow safety precautions in constructing this device. Insulate all high voltage lines properly. Use insulating box to house the circuit. The triac must be heatsinked and insulated properly.

    CAUTION!
DANGER OF ELECTROCUTION!
EXTREME SHOCK HAZARD!
THE CIRCUIT IS CONNECTED TO AC POWER LINE!

House & Car

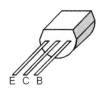

2SC3622
2SC3245
2SC3248

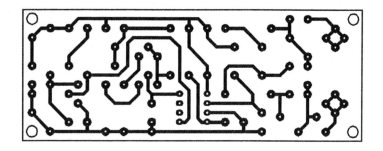

Figure 29.0 Printed Circuit Layout

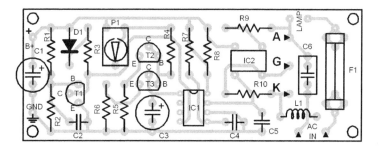

Figure 29.1 Parts Placement Layout

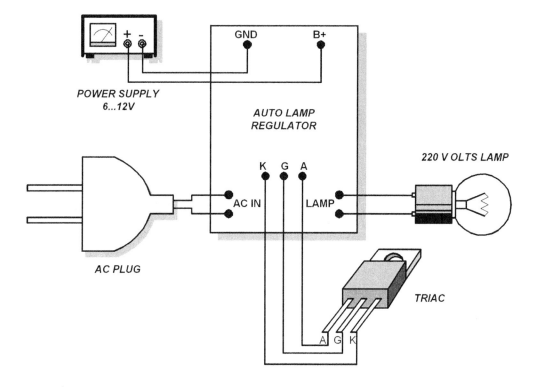

Figure 29.2 External Wirings

Electronic Circuits - 1.0

30   Universal Timer Circuit

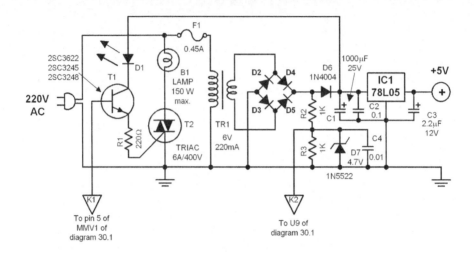

Diagram 30.0   Universal Timer Circuit (Power Supply Unit)

The time interval of this circuit can be varied digitally through the DIP switches. The time code however must be set in BCD form. A 120 Hz signal generated by doubling the line frequency is used as the time reference. This is then divided to 1 pulse per minute by the ICs 1 and 2. The counter IC3 counts backwards- that means it starts to count from the value set by the DIP switches going back to zero.

Pressing the start switch starts the counter and at the same time the triac is triggered into conduction. Once the counter reaches zero, the triac is switched off and a tone is generated. The stop button offers the possibility of forcibly stopping the counter in the middle of a counting process.

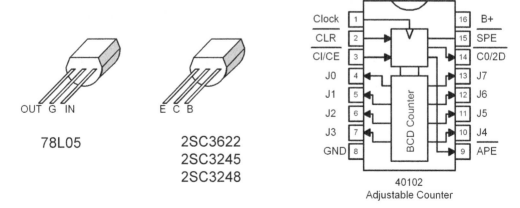

House & Car

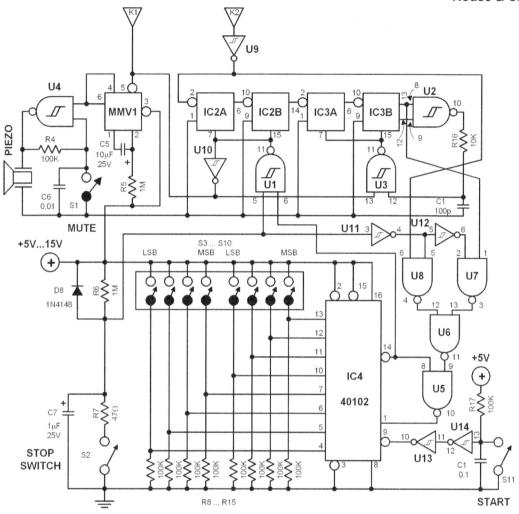

Diagram 30.1 Universal Timer Circuit (Main Unit)

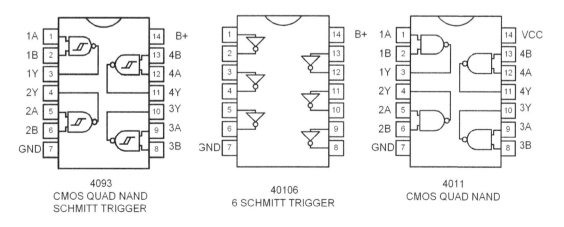

4093
CMOS QUAD NAND
SCHMITT TRIGGER

40106
6 SCHMITT TRIGGER

4011
CMOS QUAD NAND

31 Nonradioactive Smoke Alarm

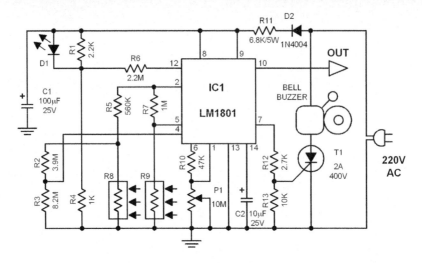

Diagram 31.0   Nonradioactive Smoke Alarm

The smoke alarm featured here uses the optical technique to detect the presence of smoke. With this technique, the use of an ionization chamber with radioactive elements is avoided. To make the entire circuit simple to construct, a single IC LM1801 is used. Take note that the circuit is directly connected to the power line, so be careful.

The actual smoke sensor is the combination of LED D2 and the two LDRs. The LED shines on the two LDRs. When a smoke appears between the LED and LDRs, the balance of the bridge circuit inside the IC changes. The IC then triggers the triac into conduction which in turn activates any alarm device connected to it. The sensitivity can be adjusted through P1.

The LDRs and the LED must be constructed inside a lightproof box. They must be installed in such a way that the smoke can pass between them.

32 Electronic Selector Switch

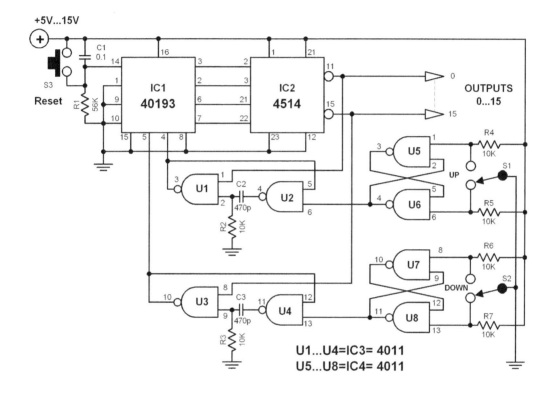

Diagram 32.0 Electronic Selector Switch

This is the electronic version of a rotary selector switch with 16 positions. It has two pushbuttons that determine whether the selector switches upward or downwards. The outputs 0 to 15 can be connected to additional driver circuits to enable it to switch higher current loads.

Switch S2 toggles the switch down while S1 toggles the switch up.

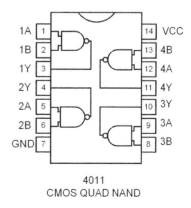

4011
CMOS QUAD NAND

Electronic Circuits - 1.0

33 Infrared Switch

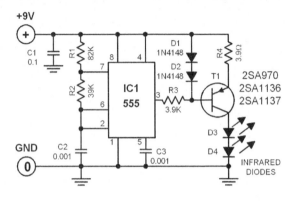

Diagram 33.0   Infrared Switch   (Transmitter Module)

Are you fascinated by automatic doors that opens by itself everytime you come close to it. Or don't you know yet how that particular door knows or senses the proximity of a moving object? Or perhaps you know the electronic principle behind it and wants to build the circuit. Most probably, the door has a special infrared switch module that is activated when the moving object interrupts its infrared beam. Such a module is very interesting to construct. It can control many types of electrical or electronic devices not just doors. A working system consists basically of a transmitter module which radiates the infrared beam, and a receiver module which captures this beam. Everytime this beam is interrupted, the receiver circuit is activated.

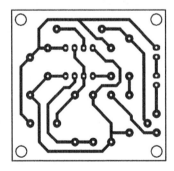

Figure 33.0.0 Printed Circuit Layout
(Transmitter Module)

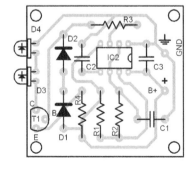

Figure 33.0.1 Parts Placement Layout
(Transmitter Module)

House & Car

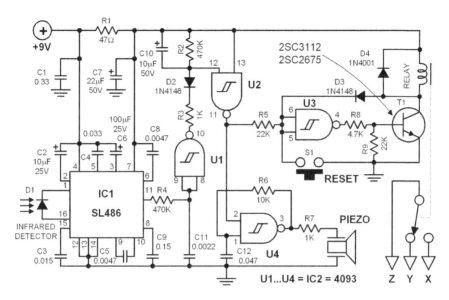

Diagram 33.1 Infrared Switch (Receiver Module)

The infrared (IR) switch featured here can be used as monitor in alarm installations, sensor in sports or optoelectronic switch in counter installations, etc. It activates when the IR beam is interrupted. It is made of two parts: the transmitter module and the receiver module. Circuit 33.0 shows the transmitter module. It is basically a squarewave generator producing a frequency of 10 kHz. The duty cycle is around 1:3. This is then converted by the infrared diodes into a pulsating infrared beam.

In normal situations, the IR beam is being received continuously by the receiver module (diagram 33.1). This module is basically a combination of preamp and a demodulator. When the IR beam is interrupted or broken, the relay in the receiver module activates and latches. At the same time, the receiver module generates a tone. After 5 seconds, the tone will stop but the relay will remain in its latched position even if the IR beam is restored. The relay will release only when the reset button S1 is pressed.

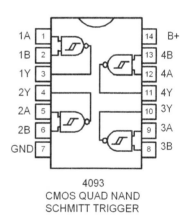

4093
CMOS QUAD NAND
SCHMITT TRIGGER

E C B

2SA970
2SA1136
2SA1137
2SC3112
2SC2675

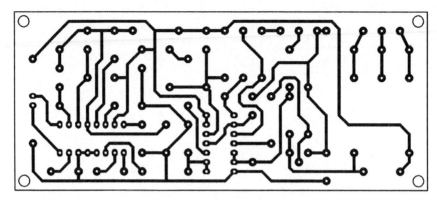

Figure 33.1.0 Printed Circuit Layout (Receiver Module)

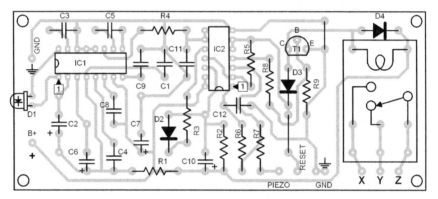

Figure 33.1.1 Parts Placement Layout (Receiver Module)

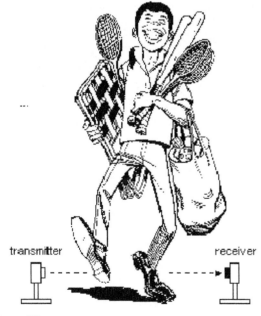

Figures 33.1.0 and 33.1.1 show the printed circuit board for the receiver module. The receiver module must be housed in a light proofed case. Only the infrared sensor D1 must be exposed in such a way that the infrared beam coming from the transmitter hits it.

Figure 33.1.2
Place the transmitter and receiver modules in such a way that the infrared beam will be interrupted by the object that is to be detected.

34 DC Flourescent Lamp

House & Car

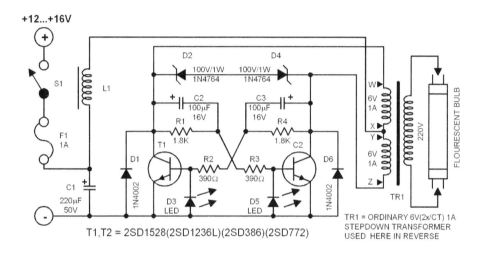

Diagram 34.0 DC Flourescent Lamp

Normally, flourescent lamps are powered by ac currents since only ac current can excite the gas inside it. If one attempts to connect a dc current to its electrodes, it will not produce flourescent light. In areas where there are no ac power lines available and the only source of electric energy is a battery, the only way to force a flourescent to light is to convert the dc current from the battery into an ac current. To do it, you need this dc to ac converter circuit. This is not just a converter, this is specially designed to produce an ac frequency which can excite the lamp to produce flourescent light.

The coil L1 is an ordinary choke coil for triacs commonly found in dimmer circuits. It must have high inductance and ampere values. If the converter circuit has a dedicated battery supply, you can remove both L1 and C1.

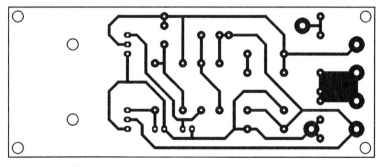

Figure 34.0 Printed Circuit Layout

Electronic Circuits - 1.0

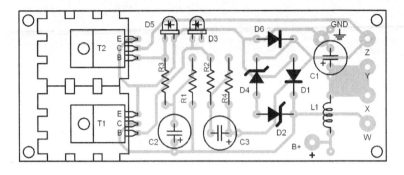
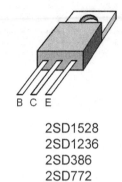

2SD1528
2SD1236
2SD386
2SD772

Figure 34.1 Parts Placement

To make the circuit easy to construct, a standard stepdown transformer is used but wired in reverse in the circuit. The transistors T1 and T2 become very hot in operation. Both must be heatsinked properly to avoid damage from overheating.

Warning: The transformer outputs a high voltage ac current. Never touch its output electrodes while the circuit is running. Insulate the contacts and wirings which carry high voltage current properly.

Parts List:

R,R4 = 1.8K
R2,R3 = 3900Ω
C1 = 220µF/50V
C2,C3 = 100µF/50V

T1,T2 = 2SD1528(2SD1236)(2SD386)(2SD772)
D1,D6 = 1N4002
D2,D4 = 1N4764 (100V/1W) Zener
D3,D5 = LED (green)
Stepdown Transformer 220V-6V(2X) center tapped
Switch (SPST) = 1 pc.
Fuse 1 A

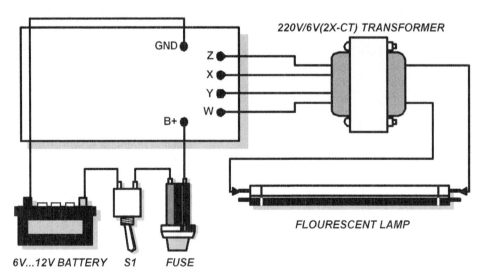

Figure 34.2 External Wirings

House & Car

35 Thermonitor

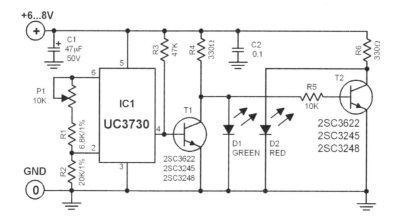

Diagram 35.0 Thermonitor

This is an electronic monitor which displays the measured value through a series of LEDs. The IC itself is the temperature sensor. It can be attached easily to heatsinks, power supplies and the likes to monitor the temperature.

When the temperature exceeds the maximum level, the green LED (D1) goes off and the red one (D2) lights up. The maximum temperature (threshold level) can be set through P1 between -1 and 100°C. The LED is a two-color type.

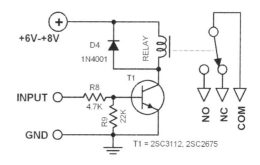

Diagram 35.1
Current amplifier with relay

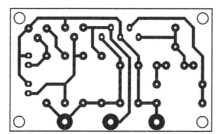

Figure 35.0 Printed Circuit Layout

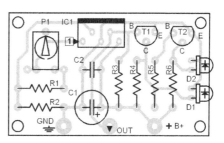

Figure 35.1 Parts Placement

Electronic Circuits - 1.0

If you use the thermonitor to control external devices like for example automatically switching on an air blower to cool down a heatsink, you have to use the current amplifier shown in diagram 35.1. This circuit boosts the trigger current coming from the pin no. 4 of UC3730 whenever the threshold temperature is exceeded. The boosted current closes the relay which in turn activates whatever external device is connected to it. Figure 35.2 shows the external wirings of the circuit's application as an automatic air blower control. The current amplifier uses the power supply of the thermonitor. This simplifies the integration of the two circuits.

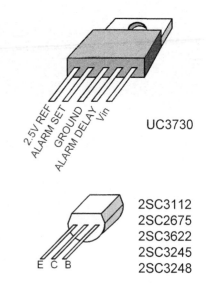

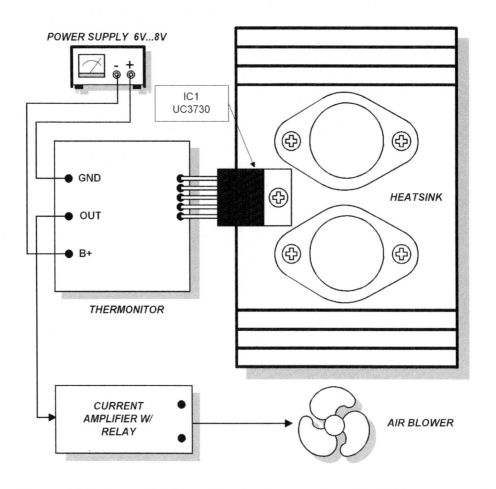

Figure 35.2 External Wirings (application as automatic air blower control)
The current amplifier w/ relay is the circuit shown in diagram 35.1

House & Car

36 Live Wire Sensor

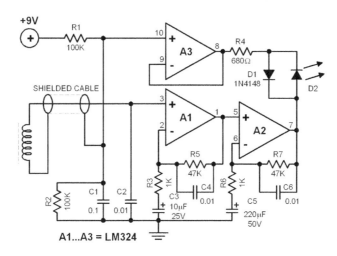

Diagram 36.0 Live Wire Sensor

This electronic device works similar to a metal detector but it does something special: it detects live electrical wires. Such a gadget is highly useful for electricians doing repair or renovation jobs. Being able to detect hidden wires will highly reduce the risk of damaging or cutting through them thus making the job safe and fast.

This circuit is specially designed for locating live wires embedded inside non-metallic walls like concrete or wood. The actual current detector used is an ordinary telephone pick-up coil. This makes the construction of the circuit much simpler compared to winding your own detector coil. The telephone pickup coil detects the magnetic field created by a live wire.

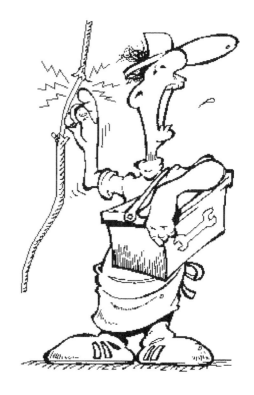

Electronic Circuits - 1.0

The detected field then induces a weak electrical signal in the pick-up coil. The circuit amplifies this signal and turns on the LED.

Using an ordinary pick-up coil saves you from constructing a sensor by yourself. This also minimizes error in constructing the circuit since pick-up coils are already tested in the factory and are guaranteed to work.

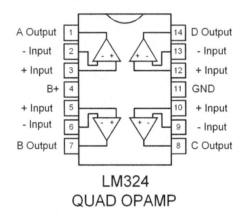

LM324
QUAD OPAMP

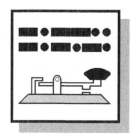

RADIO FREQUENCY

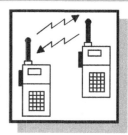

76 Diode AM Receiver
77 SSB from SW Adaptor
78 VHF Dip Meter
79 Morse Code Filter
80 VLF Converter
82 Active Impedance Converter
84 VFO Stabilizer Circuit

37 Diode AM Receiver

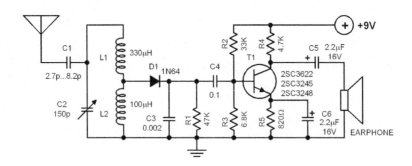

Diagram 37.0 Diode AM Receiver

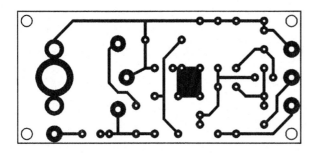

Figure 37.0 Printed Circuit Layout

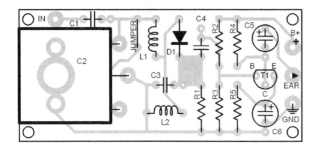

Figure 37.1 Parts Placement Layout

This is a real electronic hobbyist's project. More fun comes from building the gadget itself and less from the quality of results. The receiver circuit is a hybrid of bare essentials and a bit of technology.

It may be an "oldie" receiver but it can still provide hours of enjoyment. The original diode receiver is improved by adding an audio amplifier. The channel selection is done through capacitor C2. Coils L1 and L2 are ready made types so that you need not contruct them yourself. They are readily available from electronic supply stores. Take note of their inductance values.

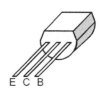

2SC3622
2SC3245
2SC3248

Radio Frequency

38 SSB from SW Adaptor

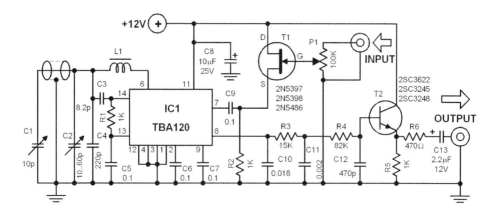

Diagram 38.0   SSB from SW Adaptor

The adaptor circuit featured here enables you to receive SSB signals with the use of a common shortwave radio.

The product detection is done by the coincidence demodulator inside the IC. The lowpass filter then improves the selectivity of the circuit. The circuit is simply connected to the IF output of the shortwave receiver and a headphone or an audio amplifier is connected to its output.

The oscillator frequency is adjusted exactly to 455 kHz with the trimmer capacitor C3. Potentiometer P1 adjusts the input level to ensure a non-distorted output signal. The circuit must be constructed inside a metal shielded box.

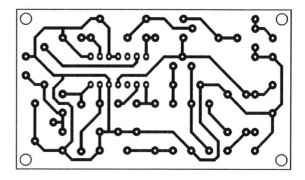

Figure 38.0   Printed Circuit Layout

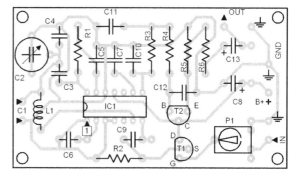

Figure 38.1   Parts Placement Layout

39 VHF Dip Meter

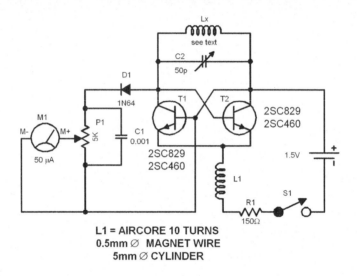

L1 = AIRCORE 10 TURNS
0.5mm ⌀ MAGNET WIRE
5mm ⌀ CYLINDER

Diagram 39.0 VHF Dip Meter

We all know how a dip meter works and what it measures - the resonance of an LC circuit. The circuit featured here is designed to work in the VHF range. The tester coil is inserted in the LX terminals and placed near the LC circuit to be tested. The variable capacitor C1 is slowly turned until a substantial dip is observed in the meter. The frequency (or the inductance) can then be read on the scale. The scale must be self constructed by using several LC circuits with known values. The fastest way to calibrate the dip meter (or to make the scale) is to use a frequency counter. The tester coil is 2 turns magnet wire. This gives a frequency range of around 50...150 MHz.

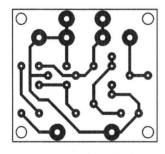

Figure 39.0 Printed Circuit Layout

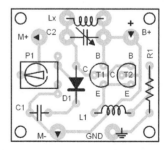

Figure 39.1 Parts Placement

Radio Frequency

40   Morse Code Filter

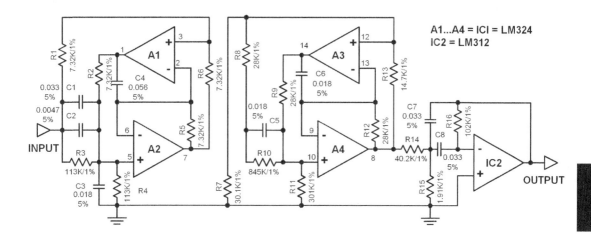

Diagram 40.0   Morse Code Filter

You certainly know the oldest encoding language for radio transmissions, don't you. You are right, it is the morse code.

Manually decoding morse code (meaning by hearing and mentally translating it into letters and words) is not highly sensitive to background noise and other signal disturbances since what we only need is to be able to differentiate the long tones (dashes) from the short tones (dots) and of course the pause between them.

Electronic Circuits - 1.0

However, if an automatic device like computer has to decode the series of tones, beeps and blips, the signal must be properly "cleaned" by a filter which removes unnecessary noise and signals. This ensures that the computer interprets the morse coded information properly. The circuit featured here is one such filter.

This morse code filter is specially designed for computer interfaces. Its center frequency is around 300 Hz. Note: A capacitor with a double value means that it is made of two capacitors connected in parallel.

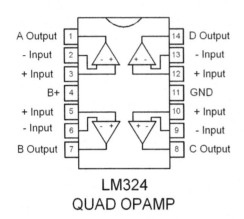

LM324
QUAD OPAMP

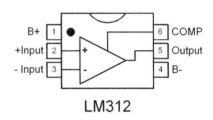

LM312

41 VLF Band Converter

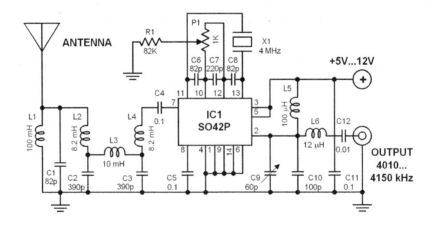

Diagram 41.0 VLF Band Converter

Radio Frequency

Did you know that there are broadcast stations below 150 kHz? Yes, indeed there are such stations and they are broadcasting in the so called very low frequency band (VLF). These are the so called low wave stations and you can hear them by using a special converter. The circuit shown above is one such converter. It converts the frequency of the low wave stations broadcasting between 10 and 150 kHz to the shortwave range of 4.01 up to 4.15 MHz. This means that you must use your shortwave radio to hear the VLF band. This technique of up-converting a signal to enable the use of a widely available receiver is very much cheaper than building a special receiver that directly processes the original signal.

Coupling the converter to the SW receiver is very simple. The output of the converter is connected in a straightforward manner to the antenna terminals of the shortwave receiver. No need to connect wires inside the radio. The channel selection is done through P1. The trimmer capacitor C1 must be adjusted to produce a maximum output. For the antenna, you can use a piece of wire. The longer the wire, the better.

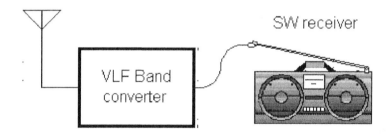

Figure 41.0 The output of the VLF Band Converter can be connected directly to the antenna of the SW receiver.

Parts List

Resistors:
R1 = 82K
P1 = 1K
L1 = 100mH
L2 = 8.2mH
L3 = 10mH
L4 = 8.2mH
L5 = 100µH
L6 = 12µH
IC1 = SO42P

Capacitors:
C1 = 82p
C2,C3 = 390p
C4,C5,C11 = 0.1
C6,C8 = 82p
C7 = 220p
C9 = 60p trimmer
C10 = 100p
C12 = 0.01
X = 4 MHz

42 Active Impedance Converter

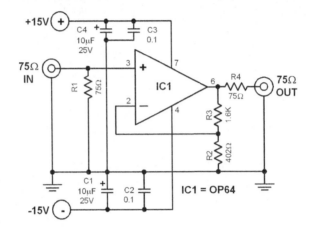

Diagram 42.0 Active Impedance Converter

Coupling two electronic modules with different impedance values together will produce sub-optimal results, unless you provide an impedance converter between the two. Simple impedance conversion can be done by arranging resistors in certain ways. However, this resistor array attenuates the signal being passed through, resulting to signal losses. To avoid losses within the impedance converter, an active circuit is usually used. An active impedance converter not only compensates for the losses, it can also amplify the signal if desired. The two circuits featured here use common OPAMP integrated circuits. They are easy to construct and use.

Diagram 42.0 shows a 75-75 ohm active converter. Since the impedances are not changed, this circuit actually works as a buffer. If dimensioned properly, it will provide some amplification.

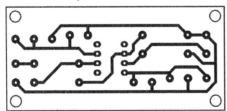

Figure 42.0.0 Printed Circuit Layout

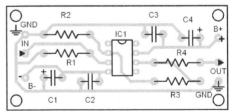

Figure 42.0.1 Parts Placement

Radio Frequency

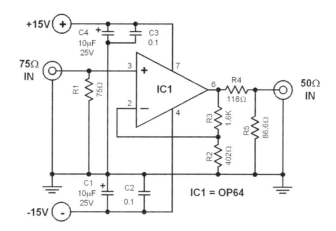

Diagram 42.1 Active Impedance Converter (75-50)

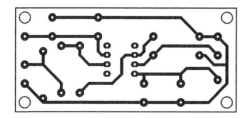

Figure 42.1.0 Printed Circuit Layout   **Figure 42.1.1** Parts Placement

Diagram 42.1 shows a 75-50 converter. It converts a 75 ohm input impedance into a 50 ohms impedance. It is usually used in pulse and video applications. It is possible to be applied as driver for coaxial cables since it can deliver currents up to 80 mA.

The first circuit amplifies the input signal by a factor of 5. However, it does not convert the impedance. The second circuit converts the 75 ohm impedance to 50 ohm impedance and at the same time amplifies the input signal by a factor of 5. The bandwidth of both circuits is 20...30 MHz.

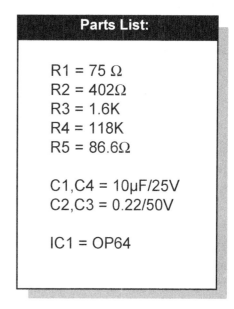

Parts List:

R1 = 75 Ω
R2 = 402Ω
R3 = 1.6K
R4 = 118K
R5 = 86.6Ω

C1,C4 = 10µF/25V
C2,C3 = 0.22/50V

IC1 = OP64

43 VFO Stabilizer Circuit

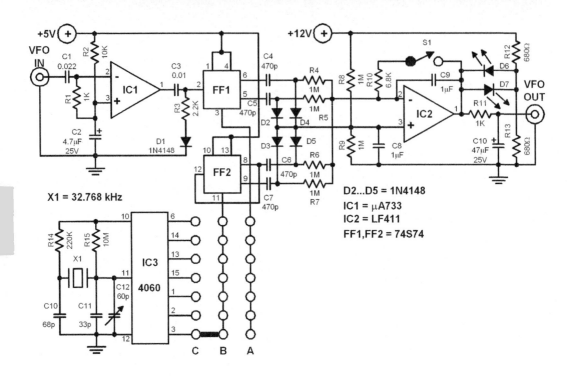

Diagram 43.0 VFO Stabilizer Circuit

Stabilizing a variable frequency oscillator (VFO) can be done by using an additional circuit when the VFO has an input through which its frequency can be controlled. Luckily, most VFOs are varicap controlled. The signal coming from the VFO is sampled and processed by the stabilizer circuit featured here. The amplified signal is then fed to flip-flop 1 (FF1). This flip-flop is controlled by a clock signal coming from 4060. The clock frequency is determined by a jumper joining points A and C. The FF1 is also controlled by another signal with a frequency which is 1/4 of the clock frequency. To achieve this, a jumper is needed between the point B and C but always one stage lower than the points C and A.

When the input signal is stabilized, the LEDs D6/D/ are off. Otherwise, a LED lights up and shows the direction of the signal shift. When the input signal is not stabilized, a difference voltage appears at the ouput which can be used as correction control for the main VFO. This VFO stabilizer circuit can operate up to 100 MHz.

HOBBY & GAMES

86 Voice Operated Switch
88 Soldering Iron Regulator
89 Electronic Pool
90 Alternating Lamps
91 Running Light
92 Projector Film Changer
94 Auto Soldering Switch
95 Bipolar Stepmotor Controller

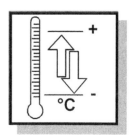

Electronic Circuits - 1.0

44 Voice Operated Switch

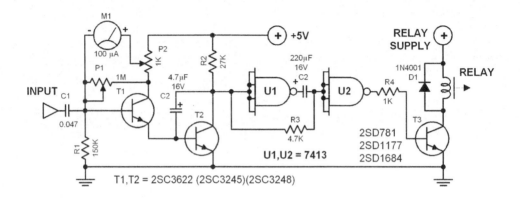

Diagram 44.0 Voice Operated Switch

There are many situations where you need to activate a certain device but you just cannot do it because your hands are busy or just cannot let go. A classic example is operating a radio transceiver while driving a motorcycle. Although you may try to drive and operate the transceiver simultaneously, this will distract and is very dangerous. The best way to solve such problem is to connect a vox (voice operated relay) to the transceiver. With the vox installed, you can remotely switch the transceiver to transmit mode by just speaking on the mic (preferably built in inside the helmet). One such vox is featured here. This relay circuit is operated by voice. The voice signal may come from a mike amplifier, an audio amplifier, or similar devices. It can be used to automatically control the transmit switch of a transceiver (VOX), a voice operated repeater, or a slide projector film changer. Obviously, the application area is not limited to motorcycling.

The sensitivity can be set through P1. Set it to a level where the vox is immune to the background noise and will only activate with your voice. The analog meter M1 is used in the circuit as a simple audio meter. Potentiometer P2 adjusts the maximum deflection of the meter. You can remove this meter if you want. However if you do so, you have to readjust the P1 to set the desired sensitivity level.

Here's a tip on one practical application of this circuit: this vox can be used as the autoswitch to activate the transmitter module of a vhf repeater system. The vox's input is simply plugged in to the receiver's audio output and its relay terminals are connected to the ptt line of the transmitter module

Hobby & Games

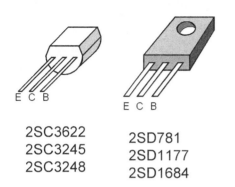

2SC3622
2SC3245
2SC3248

2SD781
2SD1177
2SD1684

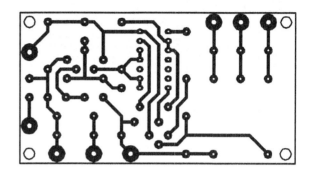

Figure 44.0 Printed Circuit Layout

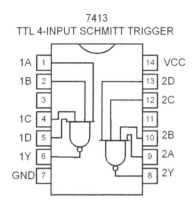

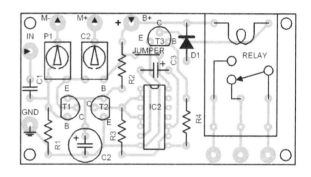

Figure 44.1 Parts Placement

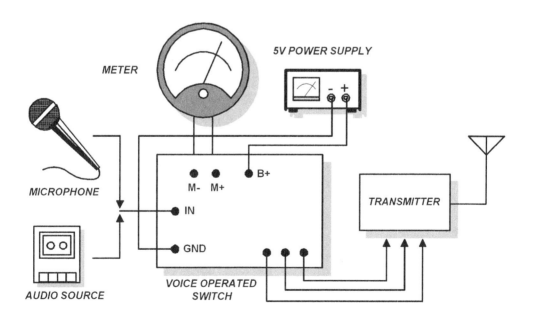

Figure 44.2 External Wirings

45 Soldering Iron Regulator

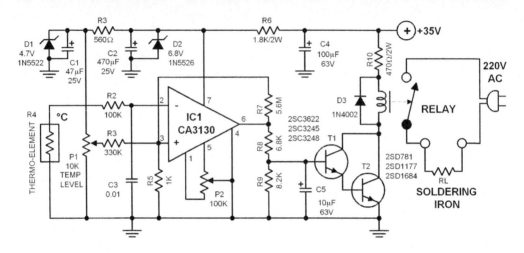

Diagram 45.0 Soldering Iron Regulator

By connecting this simple circuit to a soldering iron with a built-in thermoelement, you can easily regulate its temperature. The opamp functions as a comparator and compares the voltage value at the thermoelement with the voltage at the potentiometer P1. As long as the voltage at the minus input is lower than the voltage at the plus input, the IC is positive and T1 conducts. The relay then closes and switches on the power of the soldering iron.

The reference voltage is varied through P1 and at the same time it varies the temperature of the soldering iron. Calibration: Common thermoelements produce an output of 5 mV per 100°C. Turn P1 to the right until the relay closes and the soldering iron begins to heat up. Using a voltmeter, monitor the voltage at the minus input of the opamp and turn P2 until the relay opens when the measured voltage is 20 mV which is equivalent to 400°C.

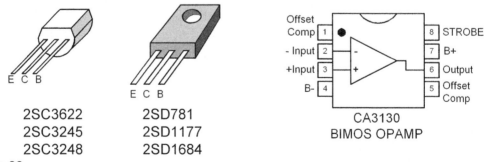

46 Electronic Pool

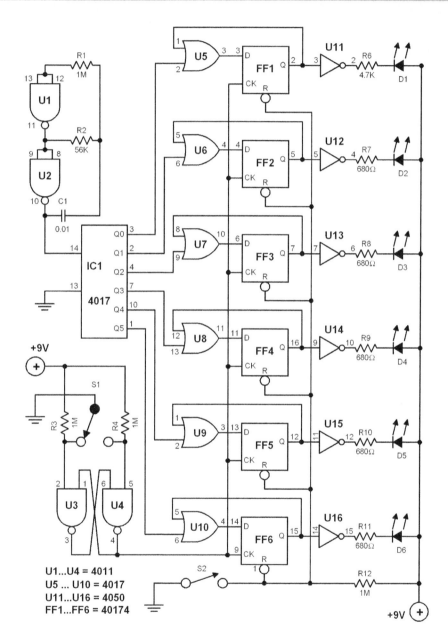

Diagram 46.0 Electronic Pool

This electronic circuit is another version of the electronic billard. The difference lies on the circuit design and the components used. Furthermore it has a "reload" switch which must be pressed before every "shot".

47 Alternating Lamps

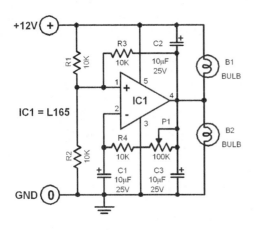

This alternating lamp signaller can be used both in model and commercial applications. The circuit can handle currents up to 3 amperes which is enough to drive two 20 watts car lamps. The IC must be attached to an adequate heatsink. The blink frequency can be varied through P1.

Diagram 47.0 Alternating Lamps

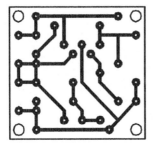

Figure 47.0 Printed Circuit Layout

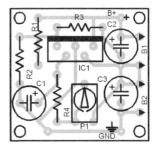

Figure 47.1 Parts Placement

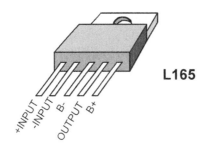

L165

48 Running Light

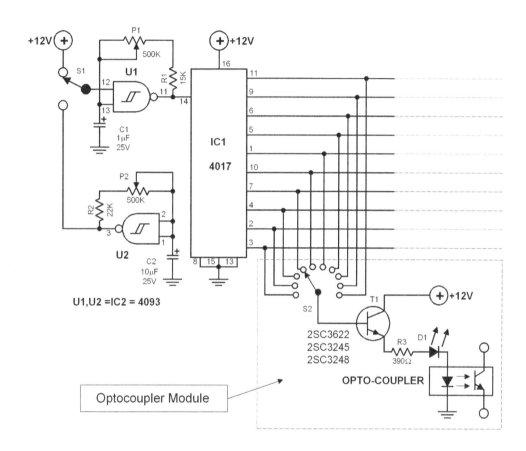

Diagram 48.0 Running Light

This running light circuit can be programmed manually to achieve the desired light combinations. Each output of the counter IC can be connected to any of the optocoupler lines through a 10-position rotary switch.

The optocoupler module must be duplicated for every lamp and connected to the output bus through a rotary switch. The running tempo can be varied through S11.

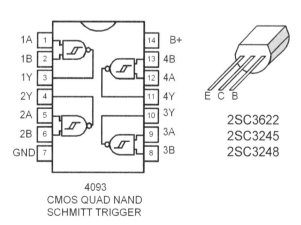

4093
CMOS QUAD NAND
SCHMITT TRIGGER

2SC3622
2SC3245
2SC3248

49 Projector Film Changer

In synchronizing the audio and the projected film, a tone code must be used. This tone code triggers a device which automatically changes the film or slide. To do this you need a two channel tape recorder. The normal audio program must be recorded in one channel and the tone code must be recorded in the other channel. The tone code channel is then connected to a decoder circuit. The decoder circuit in turn is connected to the remote film/slide changer terminals of the projector.

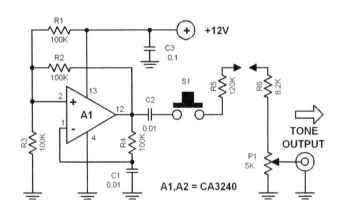

Diagram 49.0 is the tone encoder. During audio programming. The button S1 must be pressed everytime one wishes to change the film/slide so that the tone code will be recorded into the tape.

During replay, the film changes automatically being triggered by this tone code.

Diagram 49.0 Projector Film Changer (Encoder Module)

Hobby & Games

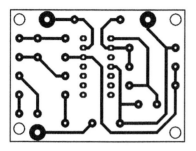

Figure 49.0.0 Printed Circuit Layout for the encoder module

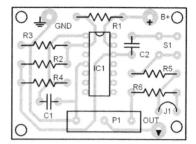

Figure 49.0.1 Parts Placement for the encoder module

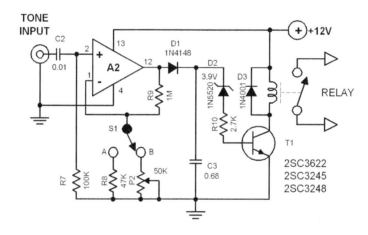

Diagram 49.1 Projector Film Changer (Decoder Module)

The second circuit (diagram 49.1) is the tone decoder. The relay contacts are connected to the remote terminals of the slide projector. The actual connections depend on the design of the projector (it can be normally open or normally closed - check your projector's handbook) . By correct programming of the tone, one can also make the film change in the reverse direction. Depending on the type of projector, it is done by either recording a single long tone or two successive short tones.

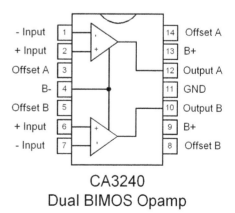

CA3240
Dual BIMOS Opamp

Electronic Circuits - 1.0

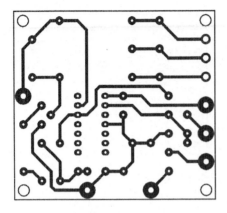

Figure 49.1.0 Printed Circuit Layout of the decoder module

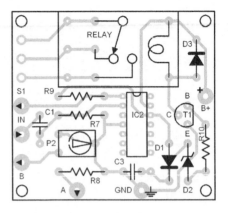

Figure 49.1.1 Parts Placement of the decoder module

50 Auto-Soldering Iron Switch

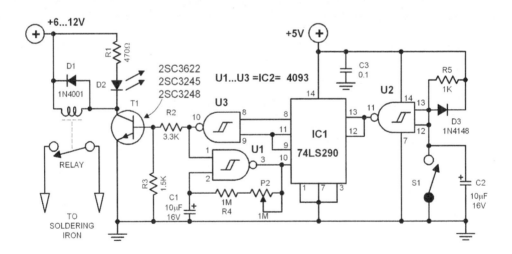

Diagram 50.0 Auto-Soldering Iron Switch

We all know that forgetting to turn off the soldering iron and leaving it for a long time on the hobby table can produce undesirable results to say the least. In order to avoid such situations and to keep the electric bill low, connect your soldering iron to this circuit which automatically turns off the power line of the soldering iron when you left it in its stand for more then 30 seconds (time allowance is adjustable through P2).

Hobby & Games

During normal soldering jobs, you must move the soldering iron every now and then to reset the timer and guarantee continued heating. This principle is similar to the deadman's switch in trains and other SPS controlled machines. The heart of the circuit is the counter IC. Its reset terminal is controlled by switch S1 which is installed in the soldering stand in such a way that it resets the counter once the soldering iron is placed in the stand. To switch on the soldering iron one must press S1 thereby closing the relay and starting the counter.

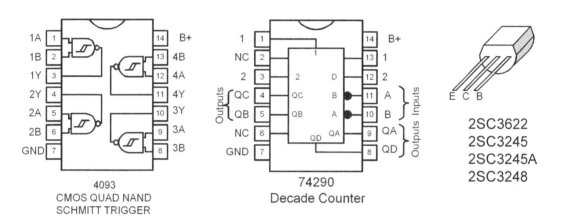

51 Bipolar Stepmotor Controller

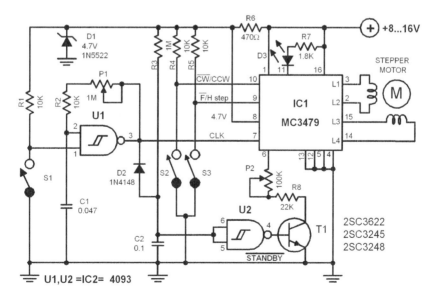

Diagram 51.0 Bipolar Stepmotor Controller

Electronic Circuits - 1.0

By using a well designed IC like MC3479, it is very easy to control a bipolar two phase step motor. The featured circuit shows how it is done. The maximum stator current can be set by the resistor R8 and P2. This resistor value is however only valid when the output transistors are not in saturated state. To find this value use the following formula:

$$I = \frac{V_s - 0.7}{R \cdot 0.86}$$

2SC3622
2SC3245
2SC3245A
2SC3248

E C B

The power supply depends on the stator coils and can be between 7.2V and 16.5V. The four ground terminals of the IC must be soldered to the ground plate. The four inputs are TTL and CMOS compatible. To control the stepmotor correctly, the following must be taken into consideration.

a) The logic at pin 10 determines the direction of the motor's rotation.
b) Pin 9 (FH) determines whether the motor moves with half or full step.
c) The clock pulse must be at least 10µS long.
d) The maximum clock frequency is 50 kHz.

The gates U1 and U2 function as a clock generator that can control the stepmotor driver stage without using a computer. Potentiometer P1 varies the output frequency of this generator and the motor's speed as a result. Of course the stepmotor can be controlled through a computer by using an appropriate program. The current consumption depends entirely on the type of the stepmotor. The IC itself consumes a maximum of 70 mA.

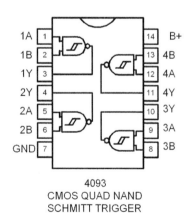

4093
CMOS QUAD NAND
SCHMITT TRIGGER

POWER SUPPLIES & CHARGERS

- **98** Polarity Protected Charger
- **100** Overvoltage Crowbar
- **102** Power Supply Regulator
- **104** PS with Dissipation Limiter
- **106** Stable Z-Voltage Source
- **107** DC to DC Converter
- **108** Versatile Power Supply
- **109** Symmetrical Auxiliary PS
- **111** Low Drop Regulator

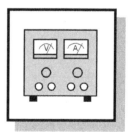

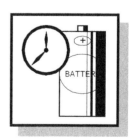

52 Polarity Protected Charger

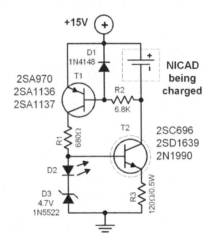

Diagram 52.0   Polarity Protected Charger

Have you ever wondered why sometimes your Nicad Battery just seem to have died after being charged. Or maybe the batteries deliver the current in the wrong direction and discharge quickly. Chances are you have placed the batteries in the wrong direction in the charger box. Charging Nicad batteries in the reversed polarity will cause no physical damage to them that is why you cannot notice your error at once. To prevent this from happening let electronics do the protection for your Nicad batteries. Use the circuit featured here. It is a Nicad charger with a built-in protector against reverse polarity charging.

This simple battery charger is designed to charge size AA Nicad batteries. It is a common knowledge that in charging batteries, the probability of interchanging the terminals' polarity is very high. If this happens, the battery might be damaged. Since we are only humans and we tend to repeat our mistakes, we better let electronics do the protection job for us and guard against this mistake. The error can never happen when using the charger circuit featured here since its charging current flows only when the battery is connected in the right direction.

2SC696
2SD1639
2N1990

2SA970
2SA1136
2SA1137

Power Supplies & Chargers

The charging current is around 50 mA. The entire circuit works as a constant current source. Polarity protection is done by T1, D1 and R1. When the battery is connected correctly, transistor T1 conducts and switches on the constant current source T2. The LED lights up when the circuit is charging. Otherwise, you know that you have made the mistake again. This time however, just reverse the batteries to start charging.

Take note: this charger circuit can charge up to maximum of 4 Nicad batteries connected in series. Do not exceed this number of batteries if you do not want to destroy the charger circuit and eventually your expensive Nicad batteries.

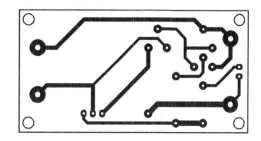

Figure 52.0   Printed Circuit Layout

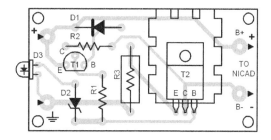

Figure 52.1   Parts Placement Layout

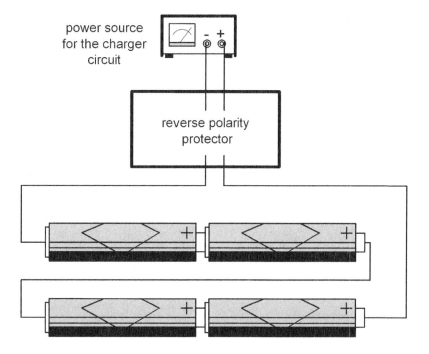

Figure 52.2   The reverse polarity protector can protect up to a maximum of 4 NiCad batteries in series.

Electronic Circuits - 1.0

53 Overvoltage Crowbar

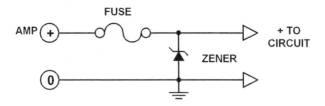

Diagram 53.0   Overvoltage Crowbar

Simple but very effective. This very simple circuit can protect a main circuit which is sensitive to overvoltages. The first circuit uses a zener diode in parallel to the power line of the protected circuit or device. Once the output of the power supply increases due to a malfunction, the excess voltage will be routed to ground by the zener diode. If the voltage will increase further, the total current consumption will exceed the capacity of the fuse. The fuse will blow at this moment and the current supply is broken.

The zener voltage of the zener diode must be 1 or 2 volts higher than the correct supply voltage and must have a higher current rating than the fuse. Of course the maximum allowable supply voltage of the protected circuit must be taken into account. For example if the supply voltage must be 15 volts and the maximum allowable voltage of the protected circuit is 18 volts, then a zener diode of 18 volts must be used.

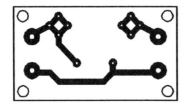
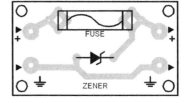

Figure 53.0.0   Printed Circuit Layout   *Figure 53.0.1*   Parts Placement

Power Supplies & Chargers

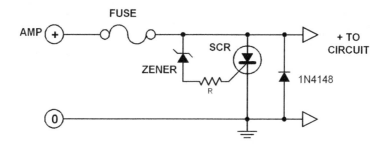

Diagram 53.1 Overvoltage Crowbar

The second circuit can be used for higher current ratings. The low current zener diode triggers the SCR when the supply voltage exceeeds the safe level. AS a result, the SCR shorts the fuse to ground thereby blowing it. The resistor R limits the trigger current and the zener current flowing through the diode. The approximate value of R can be found by using this simple formula:

$$R = \frac{\text{Zener diode's voltage} - \text{SCR's trigger voltage}}{\text{SCR's trigger current}}$$

Of course, you have to select the right SCR type for your particular application. Use the tables at the end pages of this book.

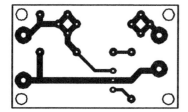 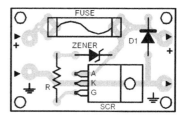

Figure 53.1.0 Printed Circuit Layout   *Figure 53.1.1* Parts Placement

Electronic Circuits - 1.0

54   Power Supply Regulator

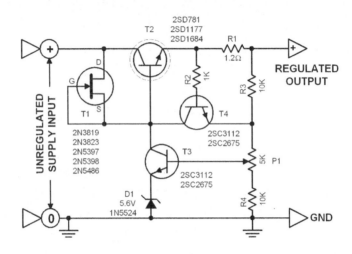

Diagram 54.0   Power Supply Regulator

Most modern power supply regulators are contructed using ICs and work more precisely. However some applications do not really require the precision of an IC. Sometimes discrete components are enough to provide the needed stability. As a hobbyists you are more interested in constructing circuits which present more challenge and chance to understand its working principles. Circuits with discrete components are more fun than ready made, ready to use ICs.

The circuit featured here is designed to give an output voltage of 12V. What is special with this circuit is that it has a current limiter circuit and a constant current source. The current output is limited up to 0.5A. The constant current source (this function is done by a single FET T1) delivers a maximum of 18 mA to the power transistor T1. The output voltage of this power supply is variable through the potentiometer P1. The potentiometer must be wired outside of the circuit board. Be careful in connecting the pot P1 to the circuit board. Never interchange the terminal connections. See Fig. 54.2 for wiring layout.

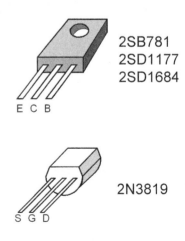

Page 102

Power Supplies & Chargers

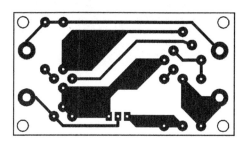

Figure 54.0   Printed Circuit Layout

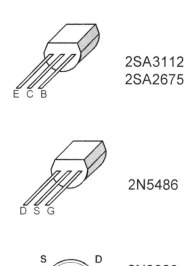

2SA3112
2SA2675

2N5486

2N3823
2N5397
2N5398

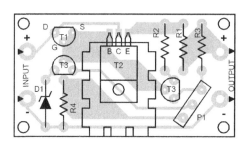

Figure 54.1   Parts Placement Layout

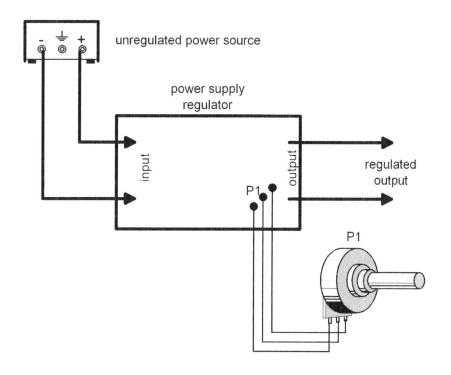

Figure 54.2 External wiring layout. Take note of the connection of P1.
Never interchange its terminal connections

Page 103

55 PS with Dissipation Limiter

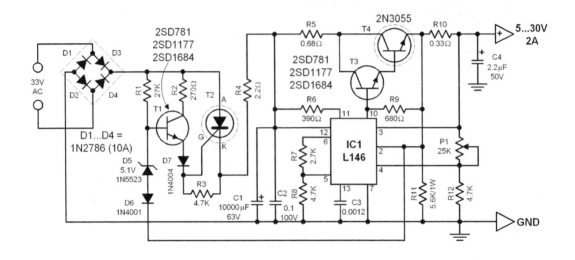

Diagram 55.0 PS with Dissipation Limiter

An ordinary power supply normally suffers from high dissipation levels. It happens when the output voltage is set at low level while the input level remains at maximum. The resulting difference voltage then "fries" the regulator circuit's power transistor continuously. This shortens the life expectancy of the transistor. Also, the heat build up in such regulators is very high that an internal air blower is almost always needed.

The power supply featured here does not know this kind of problem since it has an automatic dissipation limiter. The actual limiter circuit is composed of the components T1, T2, D5, D6, D7 between the transformer and the regulator stages. The triac T2 must be selected properly to handle the maximum current and dissipated power that will load on it. You can use the table at the end pages of this book to select the needed triac.

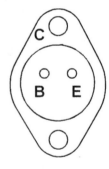

Bottom view
2N3055

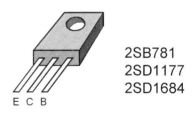

2SB781
2SD1177
2SD1684

Power Supplies & Chargers

The potentiometer P1 in the regulator stage is the normal output voltage adjuster. The output can be varied from 5 up to 50 volts.

The triac T2 and the power transistor T4 are to be installed on heatsinks outside the circuit board. The potentiometer P1 must also be installed outside the circuit board. Make sure to wire these external components correctly. Take note that two terminals of the potentiometer P1 are shorted together. See Fig. 55.2 for details.

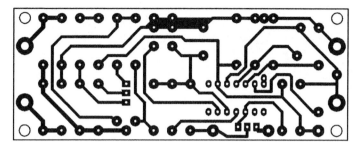

Figure 55.0   Printed Circuit Layout

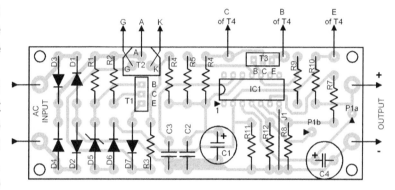

Figure 55.1   Parts Placement Layout

Figure 55.2
External wiring layout

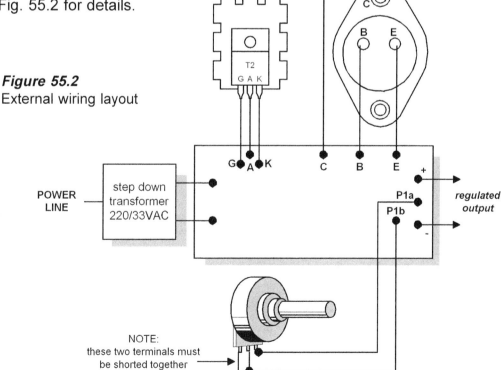

NOTE: these two terminals must be shorted together

56 Stable Z-Voltage Source

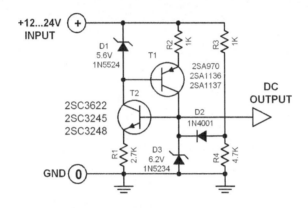

Diagram 56.0 Stable Z-Voltage Source

The zener voltage of a zener diode depends on its bias current and can vary from type to type and from capacity to capacity. It presents a problem for circuits which need a highly stable reference voltage. The solution is to make the bias current constant so that the zener voltage also remains constant.

The circuit featured here does just that. It uses a 6.2 volt zener diode. Zener diodes with different zener voltages can also be used as long as the values of R1... R4 are changed accordingly.

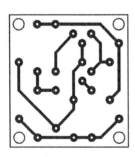

Figure 56.0
Printed Circuit Layout

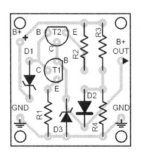

Figure 56.1
Parts Placement Layout

2SA970
2SA1136
2SA1137
2SC3622
2SC3245
2SC3248

57 DC to DC Converter

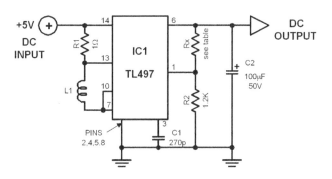

Diagram 57.0 DC to DC Converter

In some applications, one needs a voltage level that is higher than the one immediately available, like in programming EPROMS: a 25V is necessary, but common digital circuits have only 5V and 12V lines. This converter circuit solves that problem. It converts a DC voltage to a much higher level. To be able to obtain the desired output level, use Table 5.1 to know the correct value of Rx.

DC IN	DC OUT	I max	R x
5	10	125 mA	8.8
5	15	80 mA	13.8
5	20	60 mA	18.8
5	25	50 mA	23.8

Table 5.1 Value of Rx

L1 is a small coil with 85 turns of 0.2mm magnet wire around a ferrite core. Its total inductance must be around 100 µH.

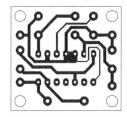

Figure 57.0
Printed Circuit Layout

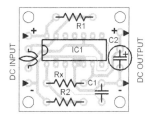

Figure 57.1
Parts Placement Layout

58 Versatile Power Supply

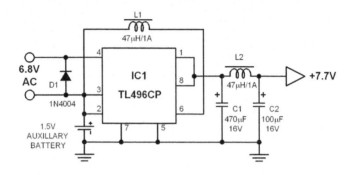

Diagram 58.0 Versatile Power Supply

This circuit is called versatile power supply. It delivers 7.7 volts output converted from either a single 1.5 volt battery or from a 6.8V AC transformer. The output is higher than the input voltage. The voltage increase is done by the integrated chip TL496CP. Not only that, both an AC input and a battery can be connected simultaneously as the power source. The AC input is controlled by a series regulator while the battery is controlled by a switching regulator.

As long as the series regulator delivers the necessary output, the switching regulator remains disabled. When a Nicad cell is used as a battery, it will be charged simultaneously by the AC input. Once the AC input is disconnected or when the AC line goes off, the battery automatically takes over the function of supplying the needed 7.7 volts. A really versatile power supply.

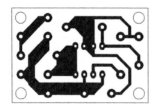

Figure 58.0
Printed Circuit Layout

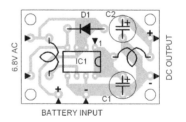

Figure 58.1
Parts Placement

Power Supplies & Chargers

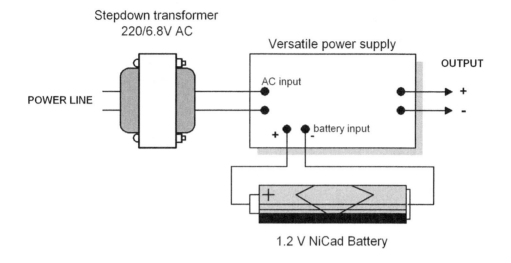

Figure 58.2 External wiring layout for the versatile power supply

59 Symmetrical Auxiliary PS

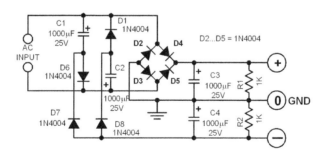

Diagram 59.0 Symmetrical Auxiliary PS

This simple circuit provides a symmetrical power supply derived from the AC output of a transformer's secondary winding without center tap. It is very useful in applications where both positive and negative lines are needed but a center tapped transformer is not available.

Page 109

Electronic Circuits - 1.0

The circuit is basically made of a bridge rectifier with additional components that function as the negative line. Take note however that this negative line can deliver less current than the positive line.

The resistors R1 and R2 are initial loads. They are added to make sure that the negative line is active even if the circuit has no load attached to its output lines. This trick is applied here because, otherwise, no negative output will be produced when the positive line has no load. In case you need the negative line to deliver more current than the positive line, just reverse the circuit.

To do this, just reverse the polarities of all diodes and electrolytic capacitors. It means all diodes and electrolytic caps, no exemption. The bridge rectifier will deliver more power in the negative line in such case.

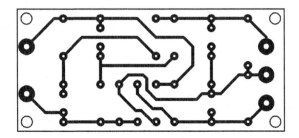

Figure 59.0 Printed Circuit Layout

The actual circuit can handle AC inputs up to a maximum of 18 volts AC. If you need a higher voltage level, change the voltage ratings of the diodes and capacitors to at least the double of the maximum AC input voltage. To select the correct diodes, see the appropriate table at the end pages of this book.

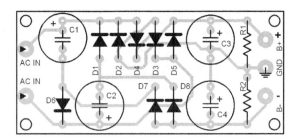

Figure 59.1 Parts Placement Layout

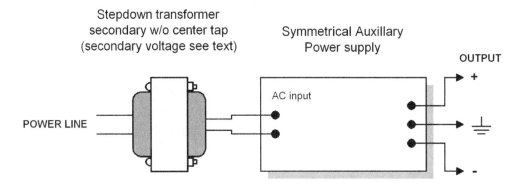

Figure 59.2 External wiring layout for the symmetrical power supply. Take note that the negative line delivers less current than the positive line.

60 Low Drop Regulator

Power Supplies & Chargers

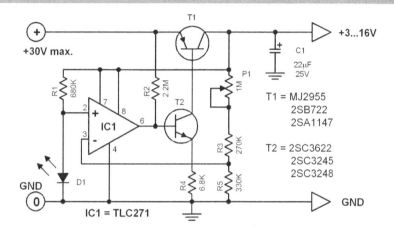

Diagram 60.0 Low Drop Regulator

2SC3622
2SC3245
2SC3248

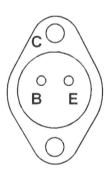

bottom view

MJ2955
2SB722
2SA1147

This power supply regulator has the following characteristics:

a) very low voltage drop of less than 1V.
b) negligible standby current from 20 to 30 µA.
c) variable but very stable output.

These characteristics make the circuit practical in some applications compared with the 3-terminal IC regulator which has high voltage drop and standby current consumption. Specially in battery operated applications, the discrete regulator featured here is highly recommended.

The circuit is basically a series regulator that uses a normal LED to produce the reference voltage. The given component values deliver a variable output from 2V up to 8V. If you intend to increase it up to 16V, change R4 with 220 kiloohms. Eventually, you need to increase the value of P1. The maximum current that can be delivered by the regulator circuit depends on the type of the series transistor used and the difference between the input and output voltages.

Electronic Circuits - 1.0

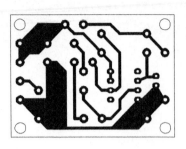

Figure 60.0 Printed Circuit Layout

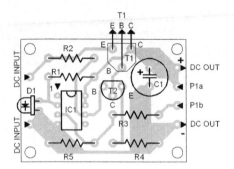

Figure 60.1 Parts Placement Layout

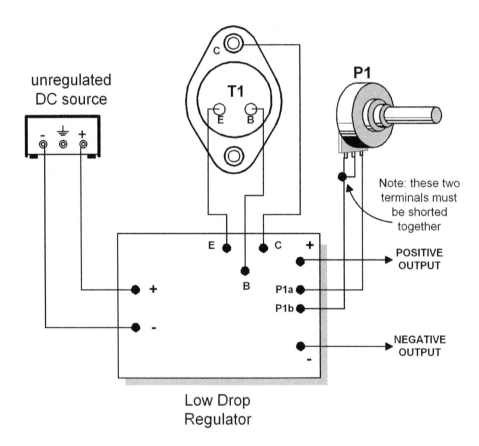

Figure 60.2 External wiring layout of the low drop regulator. Take note that two terminals of the potentiometer P1 must be shorted together. Transistor T1 must be properly heatsinked.

TESTERS & METERS

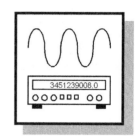

- **114** Diode Tester
- **116** Logic Probe
- **118** Poorman's Logic Analyzer
- **119** °C to Frequency Converter
- **120** Digital Ratio Meter
- **122** Digital AF Counter
- **126** LED Current Source
- **127** Wideband Signal Injector
- **129** Tendency Indicator
- **131** Wienbridge Oscillator
- **133** Cheap Frequency Counter
- **135** Light to Frequency Converter
- **136** B&W TV Pattern Generator
- **139** Acoustic Continuity Tester

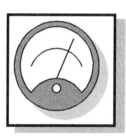

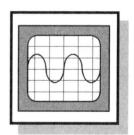

Electronic Circuits - 1.0

61 Diode Tester

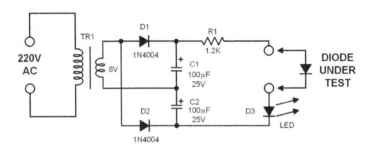

Diagram 61.0 Diode Tester

It is difficult to determine the polarity of a diode when its marking is already erased. The color ring at one end usually indicates the cathode lead. But if this color ring is erased, what are you going to do? This problem can be solved by using a multimeter. However, when one must test a large number of unmarked diodes, the multitester becomes uncomfortable to use. The best way is to build a tester circuit that is specially designed for testing diodes alone like the one featured here. This circuit not only determines the polarity of the diode but also tests whether the diode is open or short circuited.

The diode to be tested is connected to the terminals as shown. When the LED lights, the diode is connected correctly and the symbol shows the polarity. To check whether the diode is shorted, reverse the diode and if the LED remains lighted, then the diode has an internal short circuit. On the other hand when the LED does not light up in both directions, the diode is open.

The circuit can test both germanium and silicon diodes. The tester must be housed in a plastic box and the test electrodes must be marked with cathode(K) and anode(A) respectively. A box with fixed contact points installed on it is highly practical in testing a large number of diodes. A suggested design is shown in Fig. 61.3

Testers & Meters

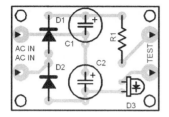

Figure 61.0 Parts Placement Layout

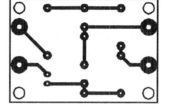

Figure 61.1 Printed Circuit Layout

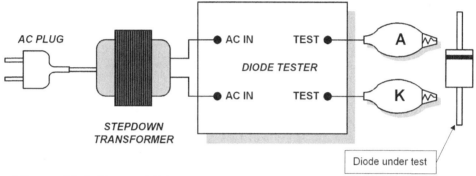

Figure 61.2 External Wirings

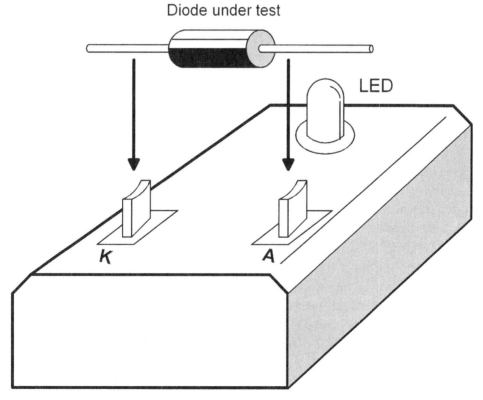

Figure 61.3 Diode tester housed in a plastic box

Page 115

62 Logic Probe

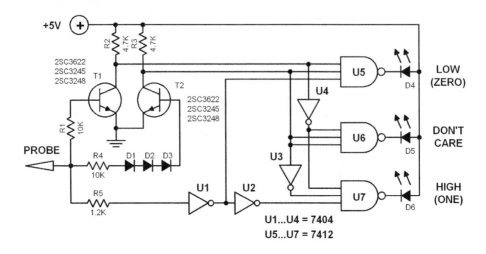

Diagram 62.0 Logic Probe

In troubleshooting digital circuits, it is always necessary to determine the logic state of a certain signal. In many cases the common multimeter device is insufficient since in the digital world, there are three logic states. Also, the technician is only interested to know the logic states but not the exact voltage levels.

This logic probe displays the three digital logic states: 1 (HIGH), 0 (LOW), and the undefined state (DONT CARE) by lighting one of its three LEDs. The logic 0 must not be higher than 0.7V and is signalled by the LED D4. The logic 1 must not be less than 2.5V and is signalled by the LED D6. The logic levels between 0.7V and 2.5V is considered as the undefined state and is signalled by the LED D5.

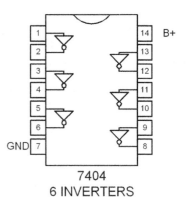

7404
6 INVERTERS

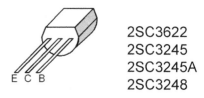

2SC3622
2SC3245
2SC3245A
2SC3248

Testers & Meters

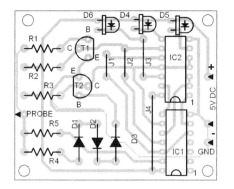

Figure 62.0 Parts Placement

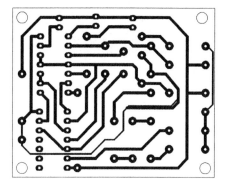

Figure 62.1 Printed Circuit Layout

Parts List:

R1, R4 = 10K
R2, R3 = 4.7K
R5 = 1.2K
U1...U4 = IC1 = 7404
U5, U6, U7 = IC2 = 7412
D1, D2, D3 = 1N4148

T1, T2 = 2SC3622
       (2SC3245)
       (2SC3248)
D4 = LED red
D5 = LED yellow
D6 = LED green

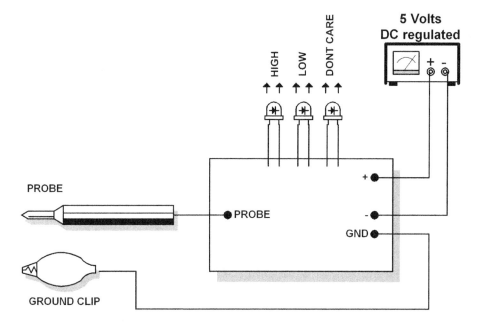

Figure 62.2 External Wirings

63 Poorman's Logic Analyzer

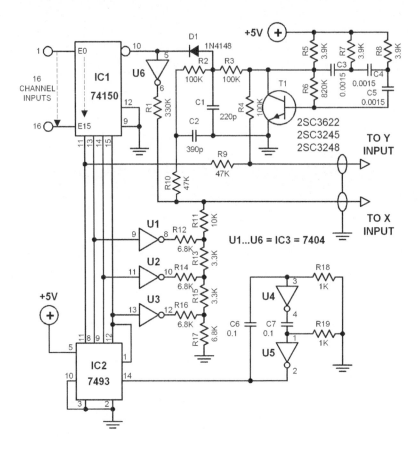

Diagram 63.0 Poorman's Logic Analyzer

This logic analyzer is called poor man's because the actual logic detector circuit is very simple and inexpensive. In order to display the logical states simultaneously, you use your own oscilloscope. The output of the analyzer circuit is connected to the X and Y inputs of the oscilloscope and the logical states are displayed on the screen in actual 1's and 0's.

Two rows of 8 logic channels appear on the screen. The circuit can monitor and display a total of 16 channels. When an input is not used, a logic 1 appears at its display position. The upper row displays the channels 1 to 8 and the lower row displays the channels 9 to 16.

64 °C to Frequency Converter

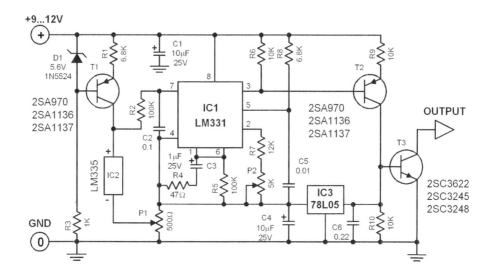

Diagram 64.0 Degree Centigrade to Frequency Converter

In some cases of temperature measurements, it is more advantageous to convert the measured value into a frequency than into a voltage. A temperature to frequency converter can be directly connected to a frequency counter or it can be connected to a computer without using any A/D interface to display the measured value. The conversion factor of the circuit featured here is 10 Hz per degree centigrade. The measurable range is from 5 to 100°C with a maximum error of (plus/minus) 0.3°C.

A single IC LM335 is used as the temperature sensor. When the temperature sensed by LM335 is zero, its output is exactly 2.73 volts. The output signal comes out in the form of a squarewave.

Calibration: Put crushed ice in a glass of water and place the temperature sensor in it. Let the sensor cool down for a few minutes. Using a voltmeter, measure the voltage level between the positive (+) pin of the temperature sensor (LM335) and pin 2 of the LM331.

2SC3622   2SC3248
2SC3245   2SA970
2SC3245A   2SA1136
           2SA1137

Electronic Circuits - 1.0

Afterwards, adjust P1 to output a voltage level of 2.73V. After this initial calibration, heat a glass of water to 50°C (use a normal thermometer to measure the temperature). Connect the display device (e.g. frequency counter) to the circuit's output. Place the sensor into the warm water, and adjust P2 to generate the desired frequency representing this temperature. For example: Adjust P2 so that the frequency is 50 Hz representing the temperature of 50°C.

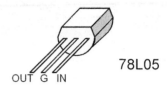
78L05

You must choose the output frequency so that you can easily adapt it to your display instrument (like frequency counter or computer).

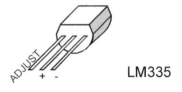
LM335

65 Digital Ratio Meter

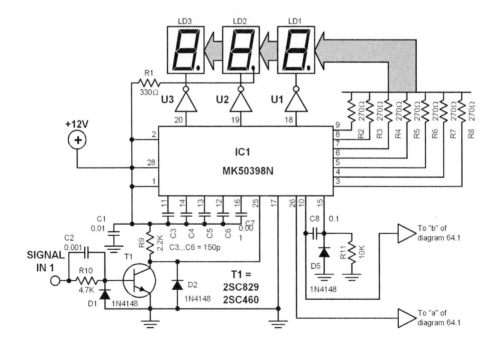

Diagram 65.0 Digital Ratio Meter

Testers & Meters

Many times, it is necessary to measure the ratio between two different frequencies, instead of measuring just the absolute value of a single frequency. The ratio meter featured here compares two frequencies, and displays the ratio between them. The higher frequency source is connected to SIGNAL IN 1 (f1) terminal, while the lower frequency source is connected to SIGNAL IN 2 (f2) terminal. The result is displayed digitally by LD1, LD2, and LD3. The measured ratio value that can be displayed is up to 99.9.

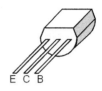

2SC829
2SC460

The heart of the circuit is a MK50398 divider IC. The displayed value is exactly the quotient of f1/f2.

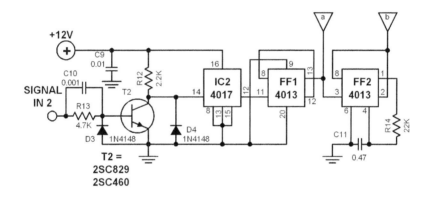

Diagram 65.1 Digital Ratio Meter (SIGNAL IN 2 front end)

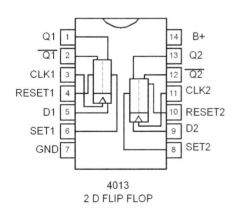

4013
2 D FLIP FLOP

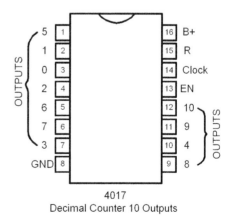

4017
Decimal Counter 10 Outputs

Page 121

66 Digital AF Counter

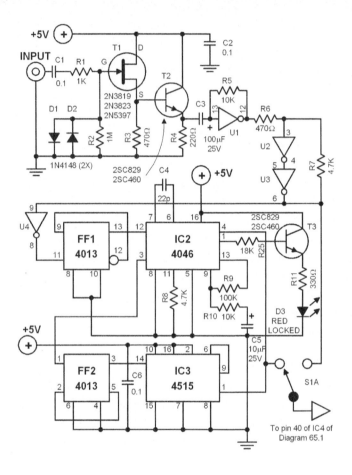

Diagram 66.0 Digital AF Counter (front end module)

A digital AF (audio frequency) counter can be constructed with only few components by using a single 7226B IC from Intensil as shown in the diagram. The upper frequency limit of this circuit is 9 MHz.

The four important functions in the frequency counter are: pulse processing, multipying, counting with timebase, and display. The input signal is converted to a squarewave pulse by the input circuitry.

Testers & Meters

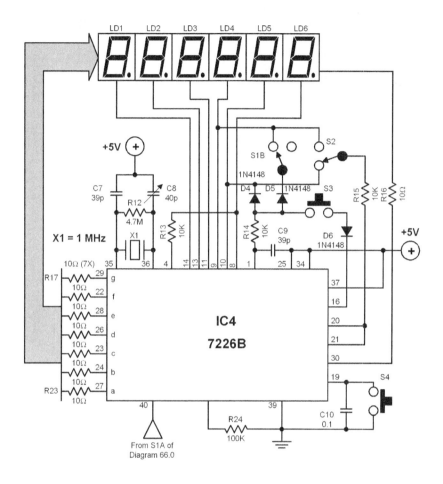

Diagram 66.1 Digital AF Counter (counter & display module)

The input circuitry is protected up to 50 Vrms. Its impedance is around 2 megaohms. The signal is then multiplied by a factor of 100. The counting process is done by the IC 7226B. This IC integrates the following circuits: oscillator with timebase; gating circuitry; counter stages; and a 7-segment display driver with multiplexer.

Switch S1 enables one to choose between the original frequency or the one multiplied by 100. Multiplying the input by 100 enables the frequency range 5 Hz to 5 kHz to be displayed. S2 selects the measuring range while S3 is the power supply switch.

The counter can be reset through S4 while S5 tests the LED display.

Electronic Circuits - 1.0

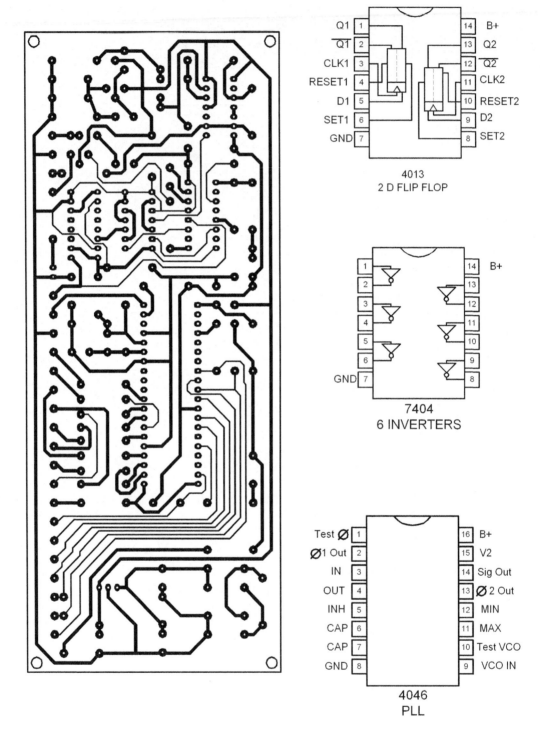

Figure 66.0 Printed Circuit Layout

Testers & Meters

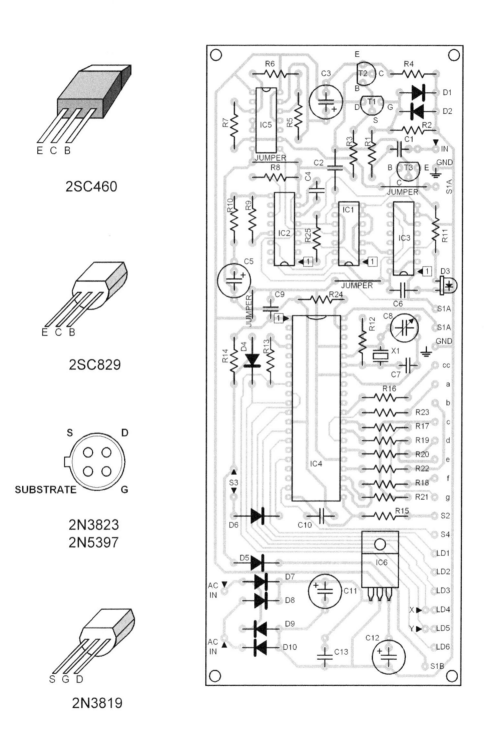

Figure 66.1 Parts Placement

67 LED Current Source

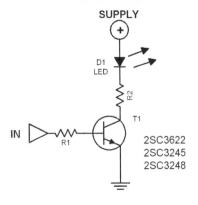

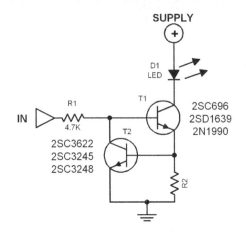

Diagram 67.0 LED Current Source

Diagram 67.1 LED Current Source

A simple LED can be used to stabilize the current in a circuit. A transistor and a LED function together as a stable current source. Circuit 1 shown in diagram 67.0 is the simplest version. T1 works as the on-off switch of the LED current source.

Circuit 2 shown in diagram 67.1 is an improved version by adding an extra transistor. The constant current delivered by the circuit can be found mathematically by: Current (I) = 0.6V/R1

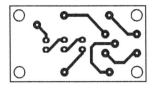

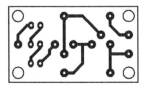

Figure 67.0.0 Printed Circuit Layout

Figure 67.1.0 Printed Circuit Layout

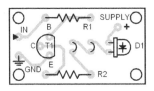

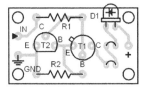

Figure 67.0.1 Parts Placement

Figure 67.1.1 Parts Placement

68 Wideband Signal Injector

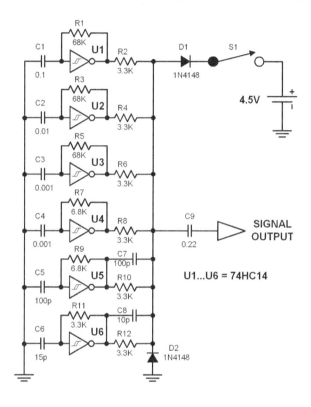

Diagram 68.0 Wideband Signal Injector

This injector circuit generates signals in both HF and AF range. This wideband capability is very helpful in troubleshooting jobs to isolate the malfunctioning stage. The circuit is constructed using 6 inverter gates inside the single IC 7414. The six gates are wired as astable multivibrators and generate a squarewave signal with different frequencies. The given component values produce the frequencies: 100 Hz, 1 kHz, 10 kHz, 100 kHz, 1 MHz, and 10 MHz.

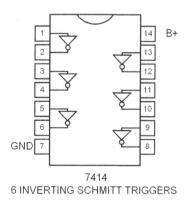

7414
6 INVERTING SCHMITT TRIGGERS

Electronic Circuits - 1.0

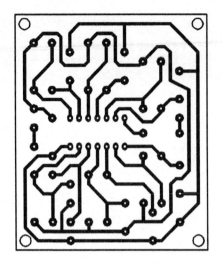
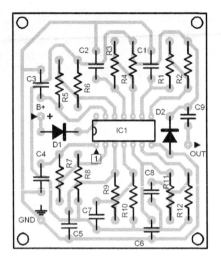

Figure 68.0 Printed Circuit Layout   **Figure 68.1** Parts Placement

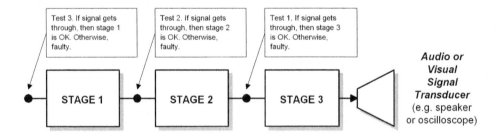

Typical troubleshooting technique using a signal injector. *You must begin at the endstage injecting the signal systematically from stage to stage going towards the input stage.*

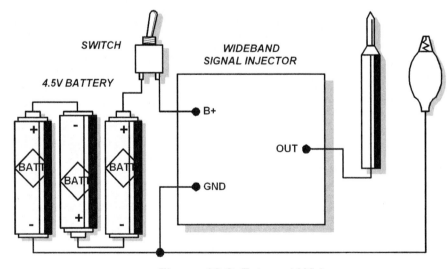

Figure 68.2 External Wiring

69 Tendency Indicator

Testers & Meters

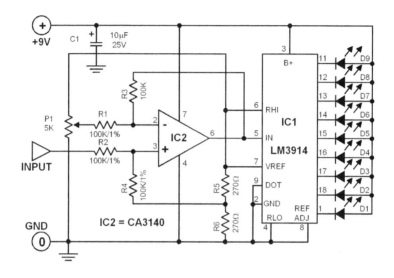

Diagram 69.0 Tendency Indicator

There are applications where a certain voltage level fluctuates and you need to constantly monitor the direction of this fluctuation. The circuit featured here displays the direction of change of a monitored voltage value. The monitored value can come from a storage battery, an electronic thermometer or a barometer, or any other power source with fluctuating output voltage. The actual monitoring is done by the differential amplifier IC2. Its output is then converted for a digital display by the IC1 LM3914. Nine LED diodes are connected to this IC and act as the level display.

The LED D5 has a color which differs from all the rest of the LEDs. It is used as the zero point - meaning the point where the monitored voltage has not changed. If the input voltage increases, one of the LEDs D6...D9 will light according to the intensity of the change. If the voltage decreases, one of the lower LEDs D4...D1 will light up. The zero point can be readjusted through potentiometer P1.

The maximum input level must not exceed 1.28 volts!

Electronic Circuits - 1.0

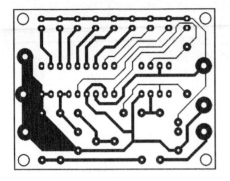

Figure 69.0 Printed Circuit Layout

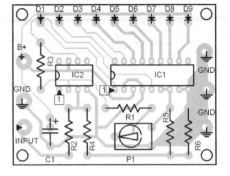

Figure 69.1 Parts Placement

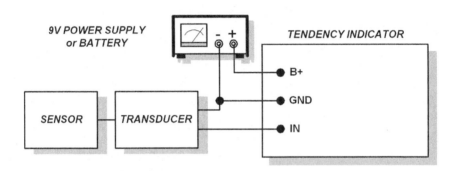

Typical installation setup of the tendency indicator. The sensor can be of any type: voltage, current, resistance, temperature, humidity, brightness, pressure, velocity, etc. The transducer is needed to convert the sensor's measured value to analog voltage value.

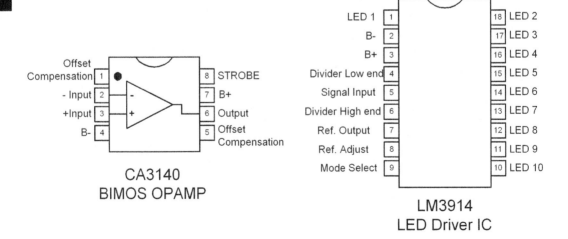

CA3140
BIMOS OPAMP

LM3914
LED Driver IC

70 Wienbridge Oscillator

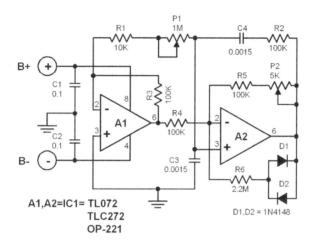

Diagram 70.0 Wienbridge Oscillator

Some electronic hobbyists are not interested in wienbridge oscillators because of the difficulty in tuning it to the desired frequency. In order to change the frequency of a wienbridge, one must adjust either two resistors or two capacitors simultaneously. This problem, however, can be avoided by using the circuit featured here. It is still a wienbridge oscillator but its frequency can be varied through a single potentiometer. The frequency (fo) can be found mathematically by:

$$f_o = \frac{1}{2\pi RC\sqrt{a}}$$

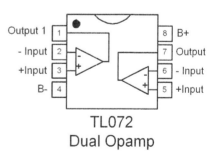

TL072 Dual Opamp

where R= R2= R3= R4= R5 and C= C3= C4 and a= (P1+R1)/R1. The maximum stable amplitude can be adjusted up to a level of 3.5V. The given component values allows the oscillator to generate frequencies between 350 Hz and 3500 Hz (3.5KHz). Potentiometer P1 varies the output frequency.

Electronic Circuits - 1.0

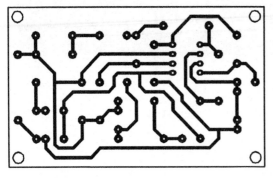
Figure 70.0 Printed Circuit Layout

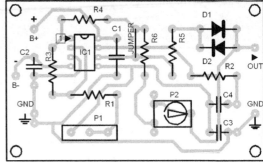
Figure 70.1 Parts Placement Layout

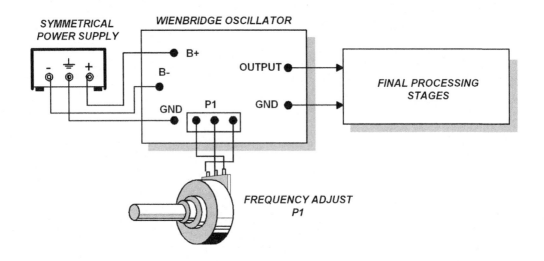
Figure 70.2 External wiring layout

Parts List:

R1 = 10K
R2,R3,R4,R5 = 100K
R6 = 2.2M

D1,D2 = 1N4148

A1, A2 = IC1 = TLC272 (TL072) (OP-22)

C1,C2 = 0.1µF
C3,C4 = 0.00015 µF
P1 = 1M
P2 = 5K

71 Cheap Frequency Counter

Testers & Meters

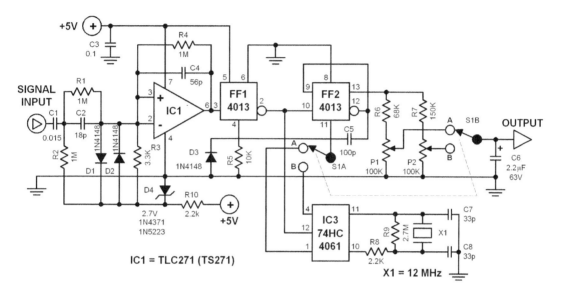

Diagram 71.0 Cheap Frequency Counter

This frequency counter uses your existing digital multimeter as the display unit that is why it can be constructed at very low cost. Due to the high impedance of most DMMs, a frequency to voltage converter can be easily connected to it without matching problems.

The converter shown in the diagram has a frequency range of 10 Hz up to 1 kHz (position A) and 1 kHz up to 100 kHz (position B). The sensitivity at low frequencies 10 Hz up to 1 kHz is around 35 mVpp and increases to 350 mVpp at a frequency of 100 kHz.

Parts List:

R1, R2, R4 = 1M
R3 = 3.3K
R5 = 10K
R6 = 68K
R7 = 150K
R8, R10 = 2.2K
R9 = 2.7M

P1, P2 = 100K trim

D1, D2 = 1N4148
D4 = 1N4371
    (1N5223)

IC1 = TLC271
    (TS271)

FF1, FF2 = IC2 = 4013

IC3 = 74HC4061

C1 = 0.015 µF
C2 = 18pF
C3, C9 = 0.1 µF
C4 = 56pF
C5 = 100pF
C6 = 2.2 µF/63V
C7, C8 = 33pF

X1 = 12 MHz Crystal

S1A, S1B = DPDT toggle switch

Electronic Circuits - 1.0

Calibration:

- First, temporarily connect the junction of R6/R7 to pin 12 of flip-flop FF2 instead of pin 13.

- Measure the voltage at C6 (the meter must be in 20V range).

- Switch S1 to position A and adjust P1 to read a voltage (at your meter) of 2.93V.

- Then, switch the meter to 2V range and adjust P2 to give a voltage of 1.875V.

- Finally, reconnect the junction of R6/R7 to pin 13 of FF2.

The DVM you are going to use must be connected to the plus and minus terminals of the converter's output. The conversion in A range is 1 kHz= 1V and in B range is 100 kHz= 1V.

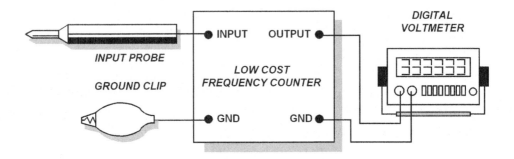

Figure 71.0 Typical installation setup of the low cost frequency counter using a digital voltmeter as the display unit.

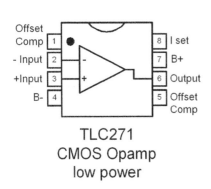

TLC271
CMOS Opamp
low power

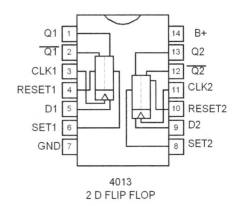

4013
2 D FLIP FLOP

Testers & Meters

72 Light to Frequency Converter

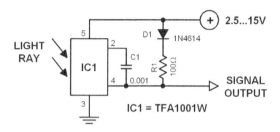

Diagram 72.0 Light to Frequency Converter

A light intensity to frequency converter with only four components? Yes it can be done by using the TFA1001W. This IC integrates a sensitive photodiode and a preamp. The IC itself functions as the light sensor. The current coming from the IC's open collector output is proportional to the intensity of the light striking the IC. The capacitor C1 causes the amplifier circuit to oscillate. The oscillation frequency is dependent on the light intensity and can be between 100 Hz and 100 kHz when the supply voltage is kept constant at 7.5 volts. Logically, when the supply voltage is of different value, the frequency range will also change. The output signal appears in a form of a squarewave. The output impedance of the circuit is quite high, that is why the circuit must be connected to succeeding stages with impedance of not less than 50 ohms.

Why convert the light's intensity to frequency? The conversion makes it easy to remotely measure the intensity level. Remote measurement is telemetry, and is very important in many applications. The signal can be channelled to a receiver through a coaxial cable. It can also be used to modulate a transmitter thus enabling the measured value to be transmitted over very long distances. If the transmitted radio signal is repeated via a satellite, the telemetry can be done worldwide.

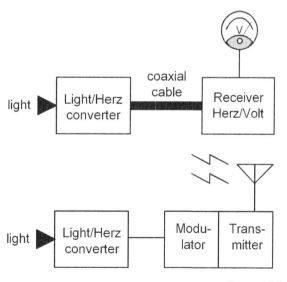

Electronic Circuits - 1.0

73   B&W TV Pattern Generator

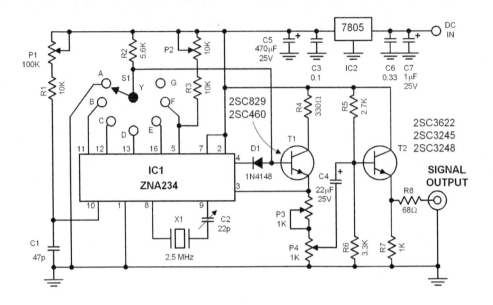

Diagram 73.0 B&W TV Pattern Generator

We all know that calibrating a televison receiver cannot be done accurately without using a pattern generator. Although adjusting the TV set using ordinary transmitted programs can be helpful, this method cannot guarantee a good alignment of the picture appearing on the screen. A pattern generator is therefore a must if you want that vertical lines be really vertical and horizontal lines be really horizontal.

The generator circuit featured here is designed for black and white TV sets. It generates seven plus two patterns. The two additional patterns are actually "no pattern" but they can, perhaps in some cases, be of help. These are: full white screen (G) and full black screen (A). The circuit itself is very easy to construct since it uses a single IC and few discrete components.

The seven main patterns are: vertical lines (B), dots (C), gray scale (D), horizontal lines (E), and checkered lines (F). Potentiometers P1 adjust the width of the vertical lines while P2 adjust the brigtness of the checkered lines. The synchronizing signal can be adjusted through P3. Potentiometer P4 varies the level of the output signal.

Testers & Meters

Patterns:
- A: Full black screen
- B: Vertical lines
- C: Dots
- D: Gray scale
- E: Horizontal lines
- F: Checkered lines
- G: Full white (blank) screen

A

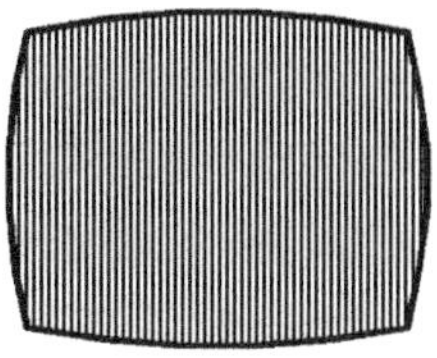

B

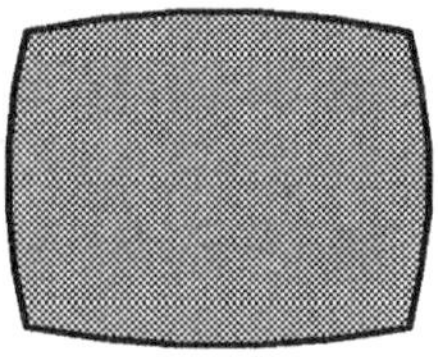

C

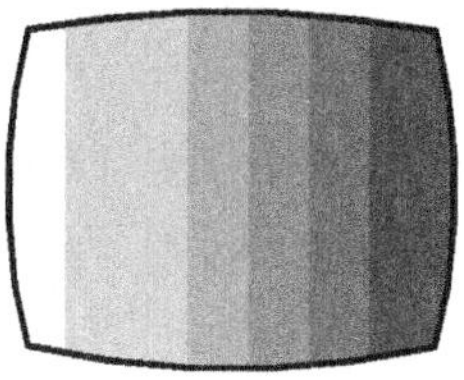

D

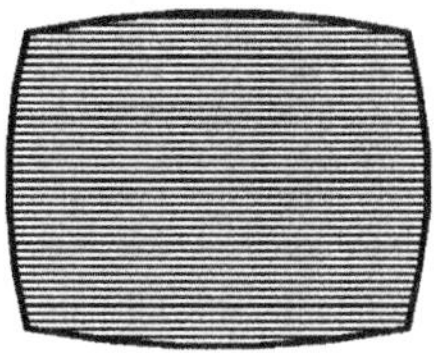

E

F

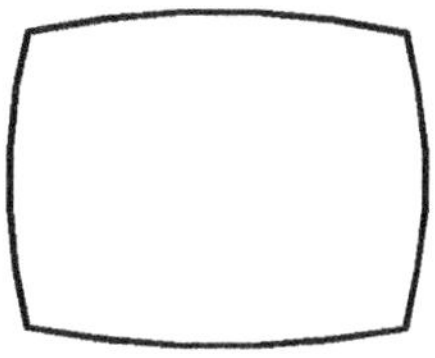

G

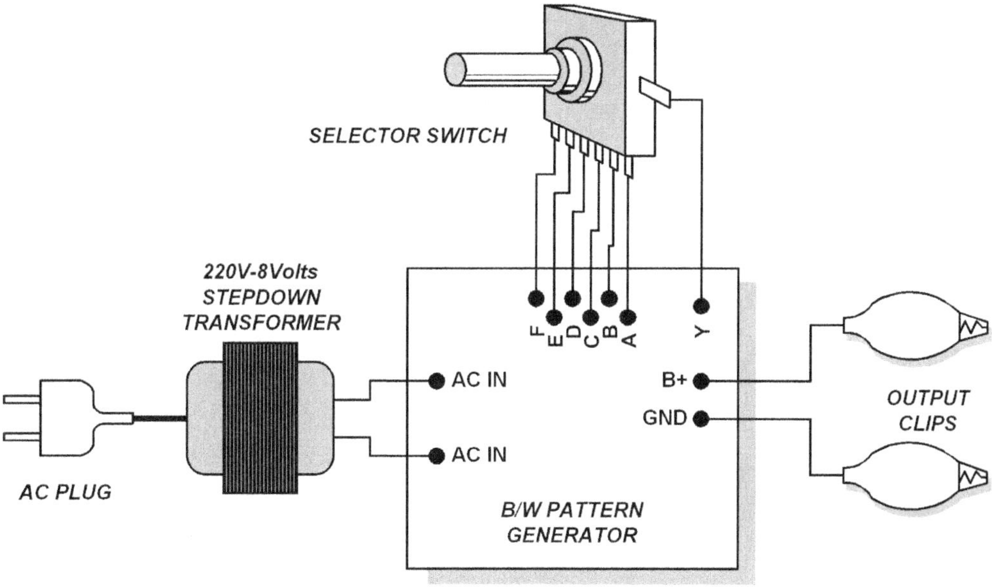

Figure 73.0 External Wiring

Electronic Circuits - 1.0

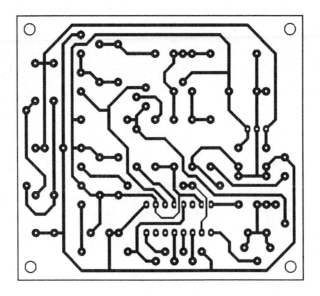

Figure 73.1 Printed Circuit Layout

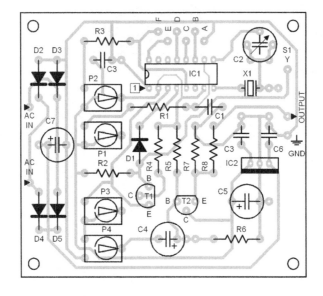

Figure 73.2 Parts Placement Layout

2SC829   2SC3622
2SC3245 2SC3245A
2SC3248

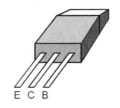

2SC460

Testers & Meters

74 Acoustic Continuity Tester

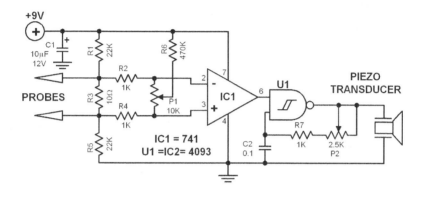

Diagram 74.0   Acoustic Continuity Tester

This tester will tell you if a contact or a line is intact by giving out a beeping tone. What is special with this circuit? First, it does not react to low resistance junctions like transistors, diodes and resistors. Second, it delivers a very low current therefore it cannot possibly damage very sensitive components like CMOS ICs.

Calibration:

Connect a 1 ohm resistor at the probes and adjust P1 until the tester beeps. Then remove the resistor and short the two probes together. In this case, the tester must also beep.

In testing contacts or wires, a resistance of above 1 ohm will prevent the circuit from beeping. The volume of beep is adjustable with P2.

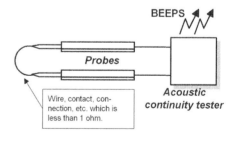

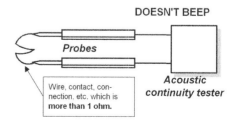

Page 139

Electronic Circuits - 1.0

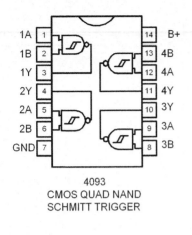

4093
CMOS QUAD NAND
SCHMITT TRIGGER

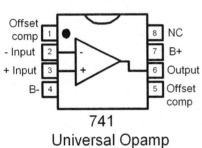

741
Universal Opamp

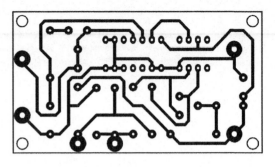

Figure 74.0 Printed Circuit Layout

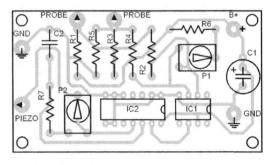

Figure 74.1 Parts Placement Layout

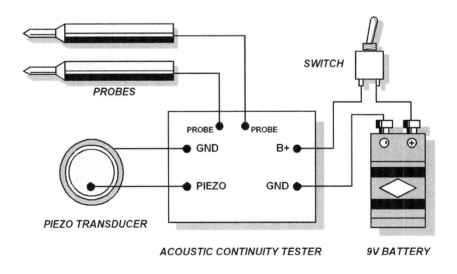

Figure 74.2 External Wirings

DIGITAL & COMPUTERS

- **142** Tape Content Monitor
- **144** Infrared Interface Circuit
- **147** Two-way RS232
- **150** Flip-flop from Inverters
- **151** Hardware Screensaver
- **153** Monitor Driver Circuit

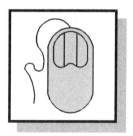

75 Tape Content Monitor

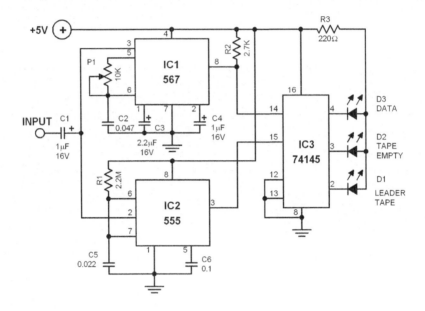

Diagram 75.0   Tape Content Monitor

This circuit displays the content of a digital data storage tape through LEDs. The LEDs are labeled as:

- **data leader** (the pilot tone at the beginning of every digital data group)
- **tape empty** (no data recorded)
- **data**

The capability to determine which part of the tape is empty and which part has a recorded data can help accelerate your work of looking for a certain program. The circuit uses a tone decoder IC 567 and a timer IC 555. The 567 IC decodes the pilot tone of each recorded program. The tone frequency to which this tone decoder reacts must be set with potentiometer P1. To do this, play a cassette with a recorded data and within the first 2 to 10 seconds try to adjust P1 so that LED D1 lights up. If you did not get it at the first try, rewind the tape and start again from the beginning.

LED D2 lights up when the tape is empty or when the playback is between two programs at the moment. When a recorded data is detected, LED D3 lights up.

Digital & Computers

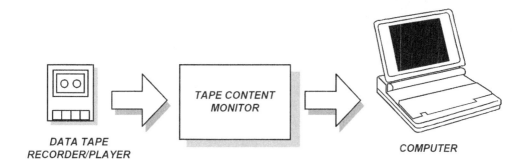

Figure 75.0 Application diagram of the tape content monitor

The input of the circuit is connected to the output of the cassette recorder. The signal from the tape reaches the inputs of the tone decoder 567 and the timer 555 simultaneously through capacitor C1. One of three things may possibly happen:

First. No signal. The tone decoder 567 outputs a logic 1 (high) and the timer 555 outputs a logic 0 (low). IC 74145 activates decimal output 2. LED D2 lights up.

Second. Leader Signal. IC 567 outputs a logic 0 (low). Timer 555 outputs logic 0 (low). IC74145 activates decimal output 1. LED D1 lights up.

Third. Data signal. IC 567 outputs a logic 1 (high). Timer 555 outputs a logic 1 (high). IC 74145 activates decimal output 3. LED D3 lights up.

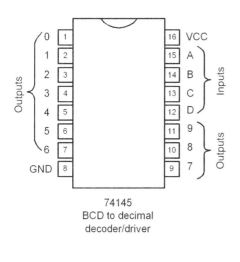

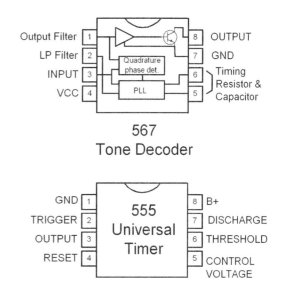

Page 143

76 Infrared Interface Circuit

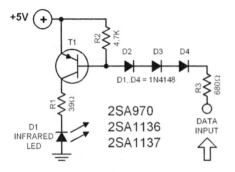

Connecting two computers together to enable data transfer between the two can be done in several ways. One method is the use of infrared beam to relay the digital data. One big advantage of the infrared signal against the radio signal is that the infrared is much less sensitive to signal disturbance.

Figure 76.0 Infrared Transmitter

The circuit featured here is a complete infrared transmitter-receiver system. It is made up of two modules. The first module is an infrared (IR) transmitter which is used as a relay in transmitting digital data. The second module receives the IR beam and demodulates the information back into digital form. This TX/RX system can be used to replace the classic cable system between computers and periphery devices. Figure 76.0 shows how the transmitter module is connected to the computer.

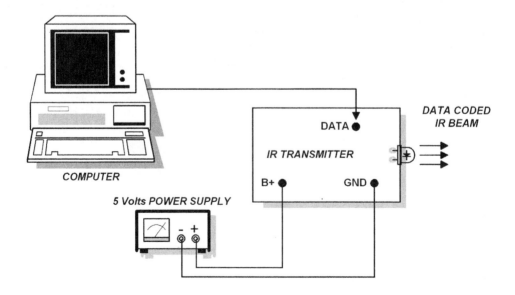

Figure 76.0.0 External wiring of the IR transmitter module

Figure 76.3 shows how the receiver module is connected to the other computer. To facilitate duplex operation, the two modules must be duplicated and connected to the computers according to the direction of the data flow.

The advantages of using cableless installation are very obvious. No more disturbing cables, freedom in arranging the different peripheries, etc. The circuit can be easily adapted to the serial I/O of a computer or a periphery device.

The potentiometer P1 in the receiver module must be adjusted for the best error-free reception.

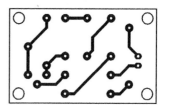

Figure 76.1
Printed Circuit Layout for the transmiter module.

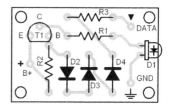

Figure 76.2
Parts Placement for the transmitter module.

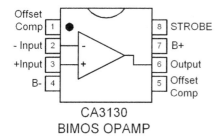

CA3130
BIMOS OPAMP

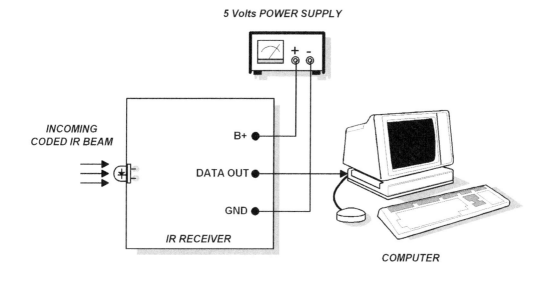

Figure 76.3 External wiring of the IR receiver module

Electronic Circuits - 1.0

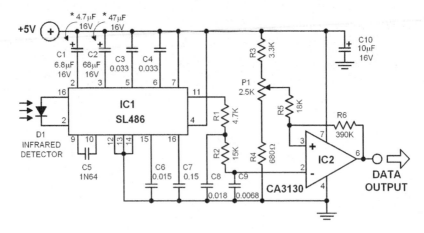

Diagram 76.1 Infrared Interface Circuit (Receiver Module)

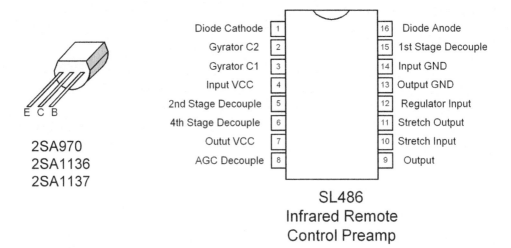

SL486
Infrared Remote
Control Preamp

2SA970
2SA1136
2SA1137

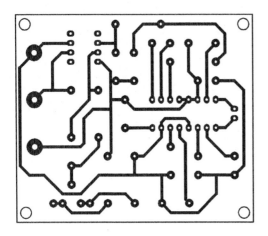

Figure 76.4 Printed Circuit Layout for the receiver module

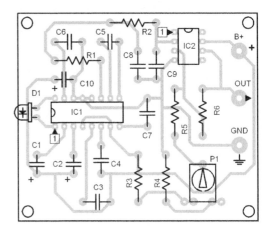

Figure 76.5 Parts Placement Layout for the receiver module

77 Two-way RS232

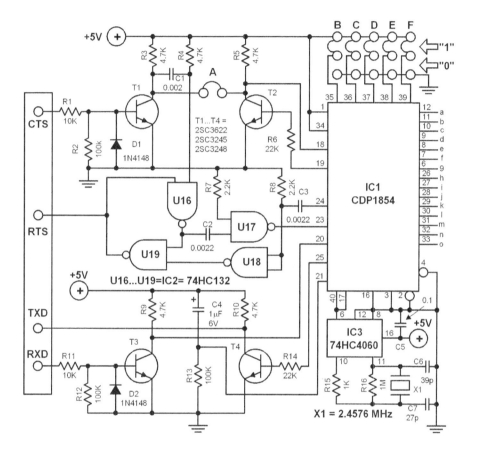

Figure 77.0 Two-way RS232

Two way means bidirectional or in other words- something can pass in both directions. This circuit is a bidirectional RS232. It is possible to transmit and receive data through it. The reason for using this device is to improve the function of the serial I/O of a computer. The circuit converts the serial port into an 8-bit parallel port (also bidirectional). This way you've got a serial to parallel and parallel to serial converter that can simultaneously receive and transmit independent from each other. It is also called Universal Asynchronous Receiver/Transmitter, or simply UART.

Electronic Circuits - 1.0

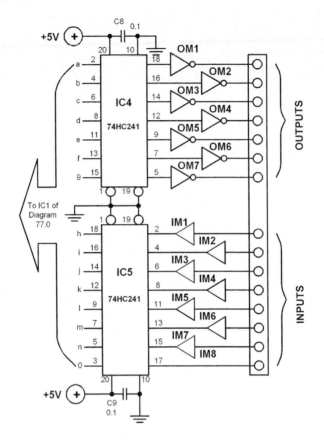

Diagram 77.1 Two-way RS232 (Input and Output)

The RT clock frequency is 19.2 kHz which generates a data transmission speed of 1200 baud. Before the circuit can be used by a computer, it must be dimensioned first by the following command:

$$COM1:30.n.8.2.CR$$

Table A shows the positions of the DIP switch for every data format. The DSR and DTR pins of an IBM PC's sub D 25 pins connector must be connected together. When no protocol is available, connect the lines RTS and CTR together.

30 means 300 baud
n means no parity bit
8 means 8 data bits
2 means 2 stop bits

Digital & Computers

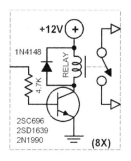

Diagram 77.2
Output module (OM1 to OM7)

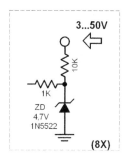

Diagram 77.3
Input module (IM1 to IM8)

bridge	connected	disconnected
A	no RTS & CTS	with RTS & CTS
B	no parity bit	parity bit
C	2 stop bits	1 stop bit
D/E	see table B	see table B
F	even parity	odd parity

Table A.

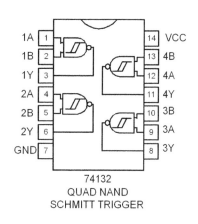

74132
QUAD NAND
SCHMITT TRIGGER

D	E	Info
0	0	5 bits
0	1	6 bits
1	0	7 bits
1	1	8 bits

Table B.

2SC696 2SD1639
2N1990

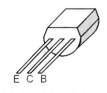

2SC3622 2SC3245
2SC3248

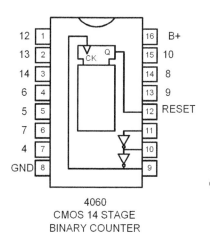

4060
CMOS 14 STAGE
BINARY COUNTER

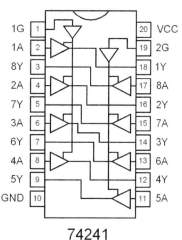

74241
8 Bus power driver

Electronic Circuits - 1.0

78 Flip-flop with Inverters

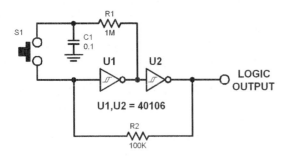

Figure 78.0 Flip-flop with Inverters

40106
6 SCHMITT TRIGGER

By constructing the circuit shown you will get a flip-flop which you can use to switch on or off any device. Of course, you must still connect the necessary driver or switch circuitry to this flip-flop.

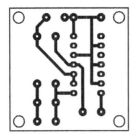

Figure 78.0
Printed Circuit Layout

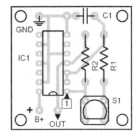

Figure 78.1
Parts Placement Layout

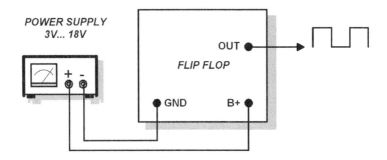

Figure 78.2 External wirings

Page 150

Digital & Computers

79 Hardware Screensaver

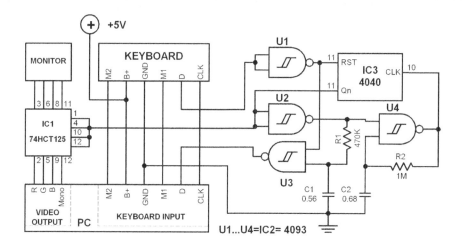

Diagram 79.0 Hardware Screensaver

Most computer fans know very well the purpose and advantage of the screensave command. It will cause the monitor to blank out when, after a preset time, nobody uses the keyboard. This is sometimes called "sleep mode". The monitor will automatically activate when a key is struck. This feature prevents the characters from being burned into the flourescent coating of the screen which happens when the computer is left operating for a long time.

Now, what is to do if your computer or software has no screensave feature? Simple, do it the hardware way! The circuit featured here blocks the RGB lines of the monitor when it detects no strobe pulse from the keyboard within a preset length of time. The circuit must therefore be connected to the monitor and keyboard lines. As you can see on the diagram 79.0, the RGB lines pass through the power bus driver IC 74HCT125. The D line from the keyboard passes through a sensor/time-out circuit composed of 4040 and 4093.

Electronic Circuits - 1.0

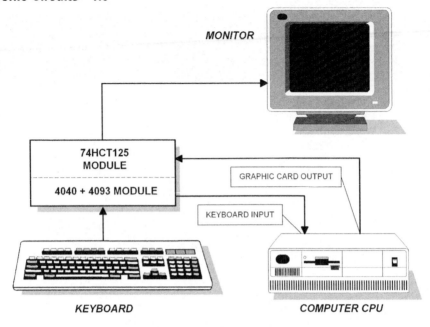

Figure 79.0 Schematic installation diagram

The D line resets this circuit when a key is struck. If, however, this circuit times out such as in the case of long standby period, the 4040 will shut off the IC 74HCT125. This IC blocks the signals that normally go to the monitor. Once the circuit detects a pulse (this pulse is generated by the keyboard everytime a key is pressed), it returns the RGB lines to their normal condition thereby displaying back the characters on the screen.

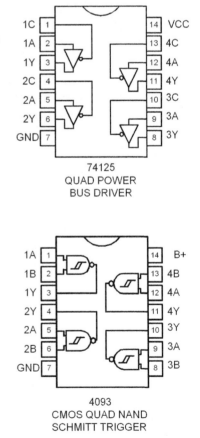

74125
QUAD POWER
BUS DRIVER

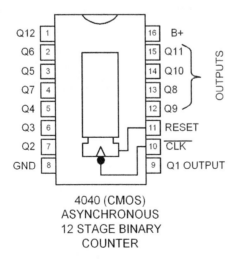

4040 (CMOS)
ASYNCHRONOUS
12 STAGE BINARY
COUNTER

4093
CMOS QUAD NAND
SCHMITT TRIGGER

Digital & Computers

80 Monitor Driver Circuit

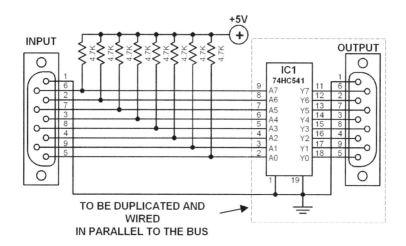

TO BE DUPLICATED AND WIRED IN PARALLEL TO THE BUS

Figure 80.0 Monitor Driver Circuit

This circuit enables you to connect four monitors to a single graphic card. The type of the graphic card used is not important (be it VGA, EGA, CGA or Hercules) as long as the logic levels are TTL compatible.

The 3-state bus driver IC 74HC541 and the multipin connector shown inside the dotted rectangle must be duplicated for each monitor and wired parallel to the main video bus. The current consumption is around 10 mA.

You must provide a regulated 5 volts to power the circuit.

Electronic Circuits - 1.0

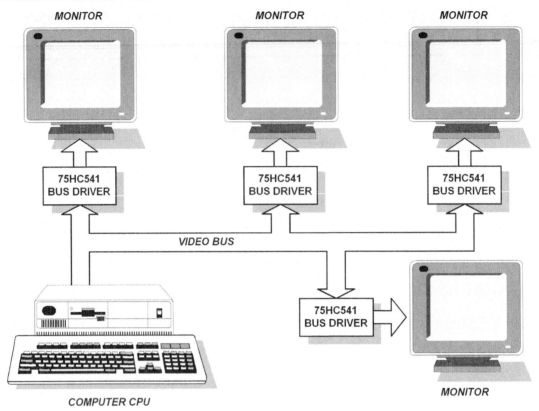

Figure 80.0 Installation diagram for four monitors

The different signals (RGB and sync) are fed to the 9-pin sub-D male connector. Their lines inside the circuit make up the main bus and each line is normally pulled up to logic 1 by a 1K resistor. The signals pass through the driver ICs which are permanently wired to conduction. They finally get to the monitor through the 9-pin sub D female connector.

Figure 80.0 shows a circuit's sample application with four monitors. You can, of course, connect more monitors. This is highly practical for information systems in large halls, in airports, transportation stations, large offices, stock exchange markets, etc.

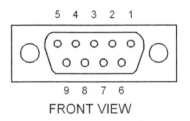

FRONT VIEW

Pin	EGA	CGA	Hercules
1	GND	CGA	GND
2	RED 2	CGA	GND
3	RED	RED	NC
4	GREEN	CGA	NC
5	BLUE	CGA	NC
6	GREEN 2	CGA	Intensity
7	BLUE 2	NC	Video
8	H-Sync	H-Sync	H-Sync
9	V-Sync	V-Sync	V-Sync

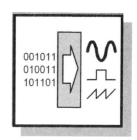

OSCILLATORS & COUNTERS

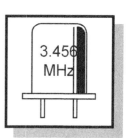

- **156** Function Generator
- **157** Duty Cycle Generator
- **158** Start/Stop Generator
- **159** Crystal Controlled Timebase
- **161** 48-MHz Clock Generator
- **163** Sine to Square/Trianglewave

81 Function Generator

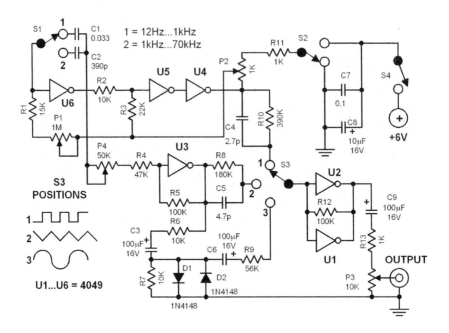

Figure 81.0 Function Generator

This is a versatile function generator that can produce three different waveforms although it is built with only one CMOS IC. The three waveforms are squarewave, triangle, and sinewave. The output frequency can be varied through potentiometer P1 from 12 Hz up to around 70 Hz in two ranges selectable through switch S1. Switch S2 selects the desired waveform.

Potentiometer P3 varies the amplitude of the output signal. Trimmer P2 adjusts the symmetry of the triangle wave while the trimmer P4 adjusts the quality of the sinewave.

Selector switch S3 selects the output waveform:

position 1 = squarewave
position 2 = trianglewave
position 3 = sinewave

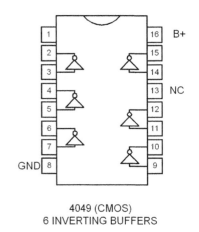

4049 (CMOS)
6 INVERTING BUFFERS

82 Duty Cycle Generator

Oscillators & Counters

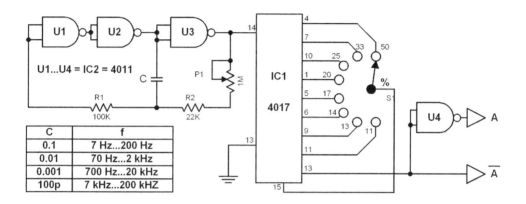

C	f
0.1	7 Hz...200 Hz
0.01	70 Hz...2 kHz
0.001	700 Hz...20 kHz
100p	7 kHz...200 kHZ

Figure 82.0 Duty Cycle Generator

A pulse generator with a variable duty cycle can be easily constructed with only two ICs. This generator is composed of a clock circuit and a decimal divider. The output of the decimal counter can be selected through a rotary switch. The diagram shows the outputs with different duty cycles. The circuit has two final outputs: A and -A(minus A).

The table shows the percentage of the duty cycle at each output. The clock frequency can be varied through potentiometer P1. If several frequency ranges are desired and you want to be able to change quickly from one range to another, connect several capacitors (for C') through a rotary switch. The capacitors' values are given in the table.

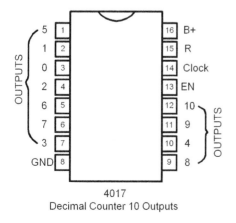

4017
Decimal Counter 10 Outputs

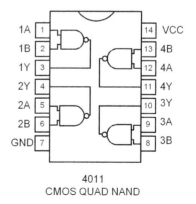

4011
CMOS QUAD NAND

Electronic Circuits - 1.0

83 Start/Stop Generator

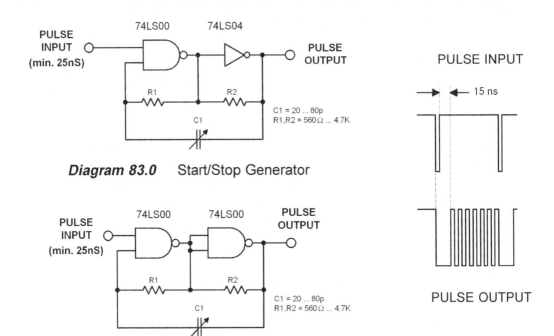

Diagram 83.0   Start/Stop Generator

Diagram 83.1   Start/Stop Generator (74LS00 version)

In applications that are using video intefaces, a device called start-stop generator is sometimes needed. Its output pulse is used to synchronize with a character pulse and creates 7 up to 12 pulses until the next character clock pulse arrives. The circuit generates its first pulse with a delay of around 15 nS so that it will not coincide with the falling edge of the input pulse.

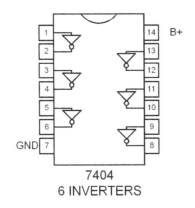

7404
6 INVERTERS

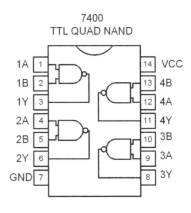

7400
TTL QUAD NAND

Oscillators & Counters

84 Crystal Controlled Timebase

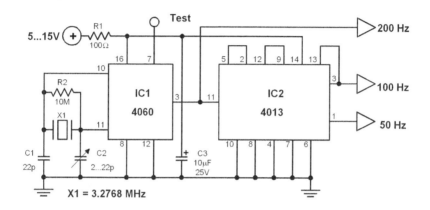

Diagram 84.0   Crystal Controlled Timebase

This circuit provides a 50 Hz timebase signal that is independent from the power line frequency. It is designed to provide the 50 Hz signal for electronic circuits which function only with this clock frequency (mostly circuits and electronic devices with european standards). It is popularly used for clock and timing applications. By carefully analysing the circuit diagram, you can see that IC1 works as crystal controlled oscillator while IC2 functions as a 2^{14} divider. The crystal frequency can be finely adjusted within 20 Hz range through the trimmer capacitor C2.

Calibration is simple and can be done in two different ways. First method: if you have a digital frequency counter, connect it to pin 7 of IC1 (TEST output on the printed circuit board) and adjust trimmer cap C2 until the frequency being displayed is 204.800 Hz.

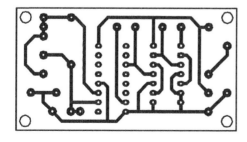

Figure 84.0   Printed Circuit Layout

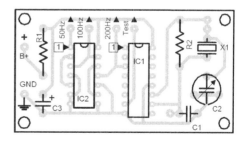

Figure 84.1   Parts Placement Layout

Electronic Circuits - 1.0

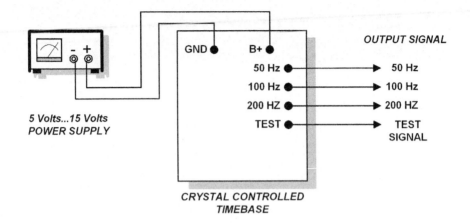

Figure 84.2 External wirings

Now, if you don't have a frequency counter; just set the trimmer capacitor C2 at its middle position. This calibration method is of course not the most accurate but will be sufficient for most applications. The output frequency, in this case, has a very negligible error of around 0.12 ppm.

This timebase source delivers three frequencies: 50Hz, 100Hz, and 200Hz. Now, if you want to divide further the output signal to obtain much lower frequencies, you can cascade additional 4013 flip-flop IC's at the output. Each 4013 has two integrated flip-flops.

You can power this circuit with voltages ranging from 5 volts up to 15 volts. This very wide power supply range is possible through the use of CMOS type ICs. The current consumption is very low - a typical character of a CMOS - between 0.4 and 3 mA.

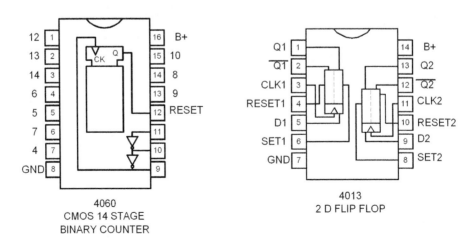

85   48-MHz Clock Generator

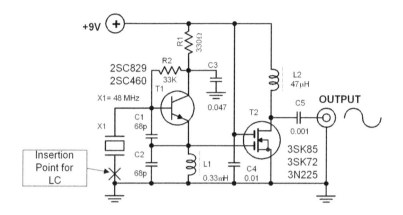

Diagram 85.0   48-MHz Clock Generator

This clock generator can be applied to serve as timebase for computers. It is crystal controlled to deliver a highly stable signal. One advantage of this circuit is that it uses a standard crystal readily available from any electronic supplier. The crystal oscillates at its third overtone between 44 and 52 MHz.

If you want to trim the crystal to be able to fine tune the frequency to exactly 48 MHz (use a frequency counter), add the LC circuit shown. The 60 pF trimmer capacitor trims the crystal frequency. The insertion point (marked with X in the circuit diagram) must be broken, and the LC circuit inserted to it in series.

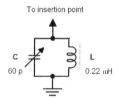

Optional LC circuit to fine tune the oscillation frequency.

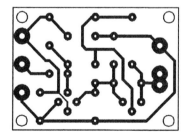

Figure 85.0   Printed Circuit Layout

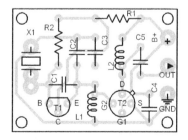

Figure 85.1   Parts Placement

Electronic Circuits - 1.0

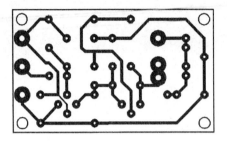

Figure 85.2 Printed Circuit Layout (with LC option)

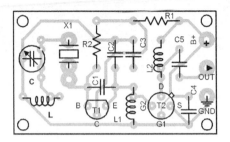

Figure 85.3 Parts placement (with LC option)

The above printed circuit layouts (Figure 85.2 and Figure 85.3) are designed for the oscillator variant with the additional LC circuit in series to its crystal. This additional LC circuit makes the oscillation frequency trimmable to an exact value.

The signal that comes out of this circuit (48 MHz) already fits into the HF category. That's why you must use a shielded cable to feed the signal to the following circuit modules that must be driven by this oscillator.

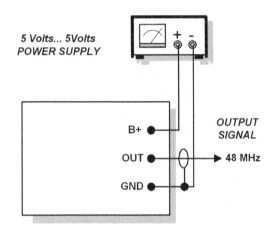

48 MHz CLOCK GENERATOR CRYSTAL CONTROLLED

Figure 85.4 External Wiring Layout

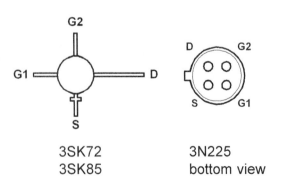

3SK72
3SK85

3N225
bottom view

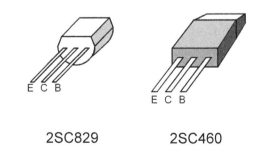

2SC829

2SC460

Page 162

86 Sine to Square/Trianglewave

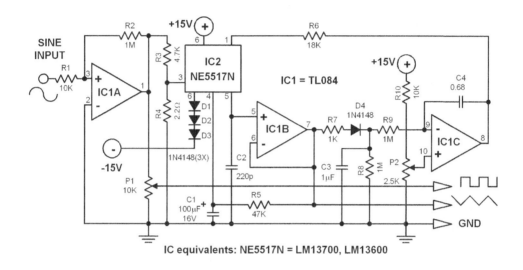

Diagram 86.0 Sine/Square/Triangle Converter

If you need a triangle wave signal or a square one but you only have a sinewave generator, use this converter to avoid the trouble of beginning from scratch. It is made up of two ICs: one LM13700 and one TL084. The amplitude of the squarewave can be varied through P1 while potentiometer P2 varies the amplitude of the trianglewave signal. The converter can process signals from 6 Hz up to 60 kHz.

In many function generators, a triangle wave generator (which is usually made of a schmitt-trigger combined with an integrator) is the heart of the circuit. The resulting sine wave is then synthesized from the triangle wave with the help of special diode networks. The circuit featured here, however, does it the other way around. First, the sinewave is reshaped to a triangle. Finally, the triangle is reshaped to a square wave.

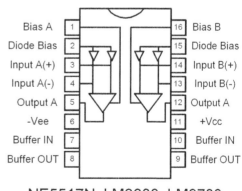

NE5517N, LM3600, LM3700
Dual Transconductance

Electronic Circuits - 1.0

The sinewave is, however, not generated by this circuit. Therefore, you must get it from a sinewave source. The sinewave is reshaped by the opamp IC1A. The two resistors R3 and R4 reduce the output level of the IC1A, since this output level swings back and forth from negative 15 volts to positive 15 volts.

The signal coming from IC1A is integrated by the combination of the transconductance amplifier IC1 and capacitor C2. The integration time factor can be easily adjusted with a preset voltage level fed to pin 1 of IC1. The opamp IC2B is added to the circuit to act as an impedance converter. It prevents voltage shifts at the capacitor C2 that might happen due to overloading.

The trianglewave can be sampled right from the output of IC2B. The following IC2C compares the trianglewave's amplitude with the value which is preset through P2. The output of IC2C controls the current regulator at the output of IC1A. This technique guarantees that the amplitude of the output signal remains independent from the input signal's frequency. The resistor R6 combined with the capacitor C1 play together so that the output level exactly matches the level that was preset by R4. They are connected in a feedback manner. This technique solves the very common problem with high precision integrators - the sensitivity to offset voltages and stray currents. For example: a stray current (maximum: 7mA at 75°C) can cause a slight change at the output signal.

We want the resulting square wave to remain clean. That is why it is important the triangle wave must remain clean. To achieve it, the time constant of the RC circuit is selected to be high. This guarantees that the triangle wave can never influence the waveform to be integrated even at the lowest frequency levels.

This converter can process signals from 5 Hz with amplitude shifts of 0% up to 60 kHz with amplitude shifts of minus 10%. At higher frequencies - above 1 kHz - the circuit needs a longer time to stabilize itself. This is due to the high time constant levels of the RC circuits.

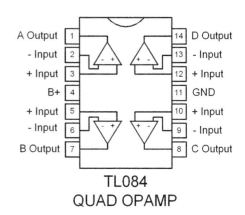

TL084
QUAD OPAMP

You must set the maximal amplitude of the source signal to 1 volt. You can also use a battery to power the circuit since the current consumption is around 10 mA only.

AUXILIARY

166 Light Activated Switch
167 Sound Generator
169 Automatic Resetter
171 DC Voltage Doubler
172 Synchronized Sawtooth
173 Running Light
175 Linear Optocoupler
176 Adjustable Zener Diode
177 Signal Light Clicker
179 Headlamp Dimmer
181 One Chip TV Audio Amplifier
182 Simple Electronic Organ
184 Voltage/Frequency Converter
186 Debounced Pulse Generator

87 Light Activated Switch

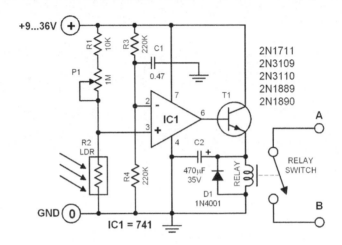

Diagram 87.0 Light Activated Switch

The brightness of the surrounding light controls the function of this circuit. The light sensor used in this circuit is a light dependent resistor(LDR). This LDR is connected to the non-inverting input of the 741 opamp IC. Once the intensity of the light striking the LDR sinks below the minimum level, the output of the opamp reverses from "low" to "high" level. The follow up transistor T1 conducts and the relay is activated thereby switching on any electrical load connected to the relay (for example: lamp). The most common application of this circuit is as an automatic switch for street lamps. Of course you can use it for other applications that require a circuit controlled by the surrounding light.

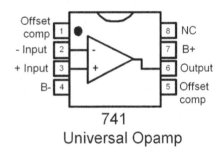

741
Universal Opamp

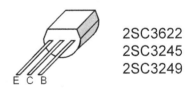

2SC3622
2SC3245
2SC3249

The supply voltage of the circuit is highly flexible. It can be between 10 and 36 volts. Potentiometer P1 adjusts the threshold level where the relay must be activated. If you intend to use this circuit to automatically control street lamps, replace P1 with 50K and R1 with 10K. The voltage rating of the relay must be selected to handle the voltage load that it is going to support.

Auxiliary

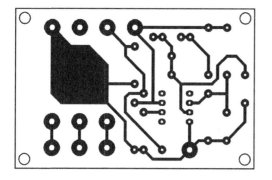

Figure 87.0 Printed Circuit Layout for the light activated switch

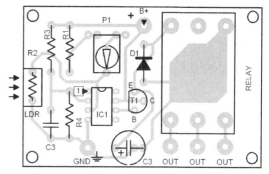

Figure 87.1 Parts Placement for the light activated switch

88 Sound Generator

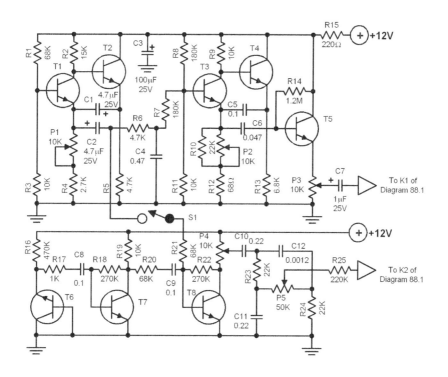

Diagram 88.0 Sound Generator

Electronic Circuits - 1.0

The construction of this quite complicated circuit will be awarded with the joy of generating almost unlimited sound effects. The circuit can produce different sounds like the heavy chopping of a helicopter's blades up to the fine chirping of a bird.

A vibrato generator (T1 and T2) modulates a tone generator (T3 and T4). The frequency of the vibrato generator can be varied through P2. A second vibrato generator is also built (T6...T10) and has the same function as the first one described.

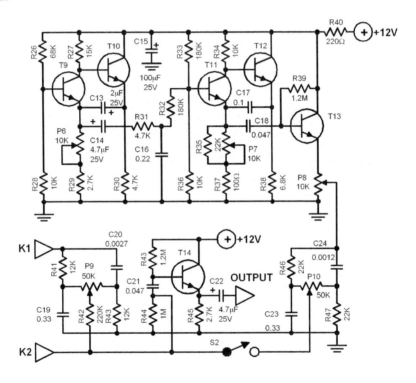

Diagram 88.1 Sound Generator

The noise sound is produced by T6 while transistors T7 and T8 amplify the generated noise sound. The potentiometers P3, P8 and P4 influence the amplitude of the generated signal and create the desired sound effect. Potentiometers P5, P9, P10 produce the desired noise envelope and strength. The mixed tone and noise signals appear at the emitter of T14. Experimenting with different potentiometer settings will certainly produce hours of enjoyment and lots of unique sound effects.

2SC3622
2SC3245
2SC3249

Auxiliary

89 Automatic Resetter

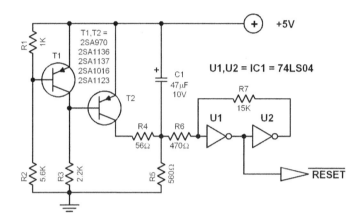

Diagram 89.0 Automatic Resetter

In some digital circuits, you must press a reset button right after you have switched on the power in order to bring the circuit to an stabilized condition. This resetting can be done automatically by this circuit. It generates a resetting pulse after the power is turned on. The reset pulse is also generated when the power has dropped to a certain low level for a certain period of time. The reset pulse is about 30 mS long.

The circuit features a very high level of immunity from short voltage disturbances appearing at the power supply lines.

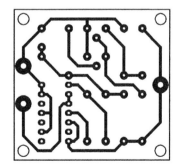

Figure 89.0 Printed Circuit Layout

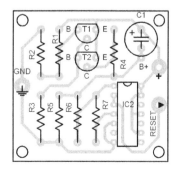

Figure 89.1 Parts Placement

Electronic Circuits - 1.0

The circuit functions in a very simple way: Right after the power supply is turned on, the voltage drop at the capacitor C1 remains at 0 volt until the power supply level reaches 4.5 volts. When this level is reached, transistor T1 begins to conduct while T2 shuts off. At this moment, the capacitor C1 charges up via R5 at a certain time constant. At this very same moment, the junction of R4, R5, and R6 drops down to zero volt.

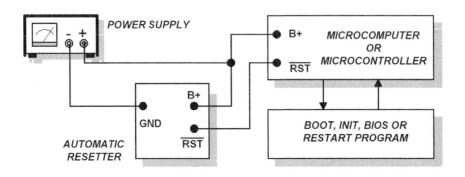

The automatic reset will cause the microcomputer/controller to automatically reload its boot or start program.

The output of U1 will become a logical 1 when a certain voltage level appears at the junction of R4, R5, and R6. Another advantage of this circuit is that it generates a reset pulse when a short power supply interruption appears. This automatic resetting is highly advantageous for microcomputer circuits which need to automatically reset themselves and load their boot or initialization program after a short power supply interruption.

2SC3622
2SC3245
2SC3249

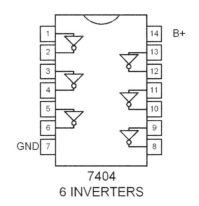

7404
6 INVERTERS

Page 170

90 DC Voltage Doubler

Doubling a voltage level is sometimes necessary to power a certain circuit. Voltage doubling can be done by the very simple circuit featured here. The circuit is controlled by a clock signal of around 10 kHz. The signal's peak amplitude must be strong enough to trigger the transistor T1 to conduction.

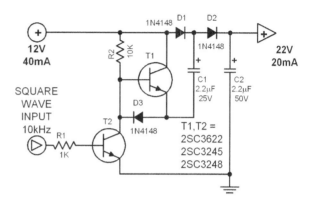

Diagram 90.0 DC Voltage Doubler

When T2 conducts, the capacitor C1 charges up to the power supply level. When the signal inverts to zero level, transistor T2 shuts off while transistor T1 conducts. At this moment, the capacitor C2 - which was initially full charged from the power supply line - will be charged some more by the series combination of capacitor C1 and the power supply line.

After some time, the capacitor C2 will be charged to a level which is almost double the power supply level. The resistance value of R1 (set to 1K in the diagram) is dependent on the input signal's amplitude. You must adjust this value to adapt it to a different amplitude level.

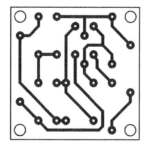

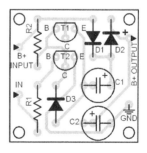

Figure 90.0 Printed Circuit Layout   **Figure 90.1** Parts Placement

91 Synchronized Sawtooth

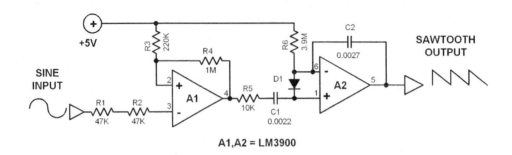

Diagram 91.0   Synchronized Sawtooth

The sawtooth signal generated by this simple circuit is synchronized with the power line frequency. It is originally designed to control triac circuits but it can be used for other special applications.

The supply voltage is highly flexible and can be any level between 4 volt and 36 volts. The values of R1 and R2 are highly dependent on the maximum input voltage. The best way to obtain the input signal is to connect the input to the power line through indirect means like using an stepdown transformer or a high voltage capacitor.

The first opamp A1 converts the power line sinewave signal into a squarewave signal. This squarewave is further fed to the second opamp which converts it to a sawtooth formed signal. The opamp A2 functions as a conventional integrator. Its output sinks linearly since a constant current flows through R6 to its non-inverting input. This integrator is reset to its initial state periodically with the help of the input signal. The internal design of the opamp makes the integrator react to both positive and negative input signal.

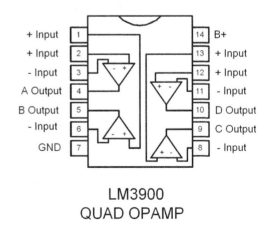

LM3900
QUAD OPAMP

Auxiliary

92 Running Light

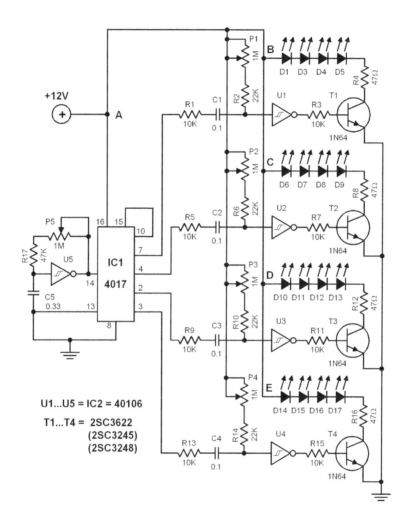

Diagram 92.0 Running Light

This circuit controls several LEDs to simulate the effect of a running light. The "running speed" is dependent on the frequency of the clock generator and can be varied with the pot P5. When P5 is in the middle position, the frequency is around 6 Hz. How long the LEDs remain lighted is controllable through the potentiometers P1 up to P4. To make the LEDs appear to "run" smoothly, set all potentiometers (P1 to P4) to the same position.

Electronic Circuits - 1.0

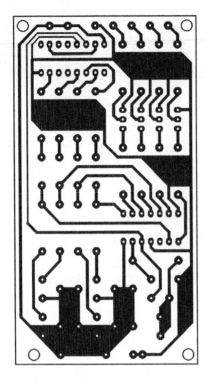

Figure 92.0 Printed Circuit Layout

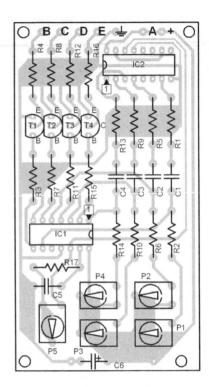

Figure 92.1 Parts Placement

Figure 92.2 below shows how you must arrange the LEDs to achieve the best effect. If you desire to use this circuit to control high current lamps or triacs, replace the LEDs with optocouplers. The internal LEDs of the optocouplers will then replace the normal LEDs. The optocouplers will act as isolated relays to trigger higher current capacity switches or relays.

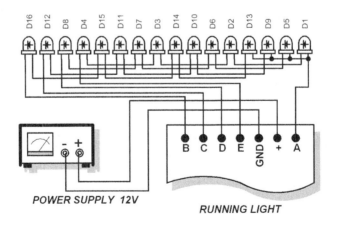

Figure 92.2 Optimal arrangement of the LEDs

93 Linear Optocoupler

Auxiliary

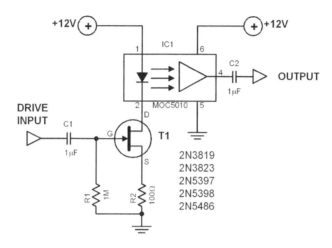

Diagram 93.0 Linear Optocoupler

Optocouplers are highly valuable in electronics as isolators in power supply circuits, audio input/output, medical applications, etc. An optocoupler like the special IC used in the featured circuit has a typical isolation resistance of 10^{11} ohms.

The principle of an optocoupler is very simple: the changes in the input current produce corresponding changes in the output current. The input circuit is, however, electrically and galvanically isolated from the output circuit. The signal transfer is done through the use of either light, laser, or infrared transmitter and receiver packaged and sealed in a single unit.

The circuit shown is an optocoupler driver and has an amplification factor of 0.75. The maximum input voltage must not exceed 2Vrms. Its bandwidth is 118 kHz. The FET in the circuit works as voltage to current converter. As is usual with optocoupler circuits, each side must have its own supply line. Otherwise, the desired high degree of isolation will not be achieved.

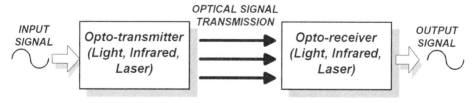

Signal transfer is done via optical transmission

94 Adjustable Zener Diode

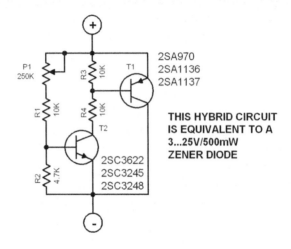

Diagram 94.0 Adjustable Zener Diode

This circuit functions as a zener diode and has the advantage of having a variable zener voltage. It is sometimes called hybrid zener diode. It is more versatile and more adaptive than fixed value zener diodes. The total resistance of this circuit is 20 up to 50 ohms and its maximum dissipation is negligible. The zener voltage is variable from 3V up to 25V with the potentiometer P1.

Make sure that the maximum current through the circuit does not exceed 100 mA !

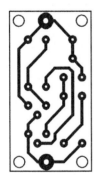

Figure 94.0
Printed Circuit

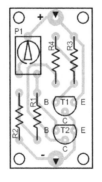

Figure 94.1
Parts Placement

2SA970   2SC3622
2SA1136   2SC3245
2SA1137   2SC3249

Auxiliary

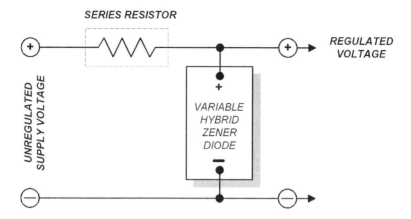

Hybrid zener diode used as a classical voltage regulator.

The figure above shows the hybrid zener diode in the most classical application of a zener diode - as a voltage regulator. The series resistor in the circuit is very important - it dissipates the excessive voltage so that the output voltage remains at a constant level. You have to use the standard procedure in determining the value of the series resistor for a zener diode voltage regulator circuit.

95 Signal Light Clicker

When the car's signal light blink, the clicking sound of its mechanical controller can be heard. This click serves as an acoustic feedback for the driver to know whether the signalling system is working or not.

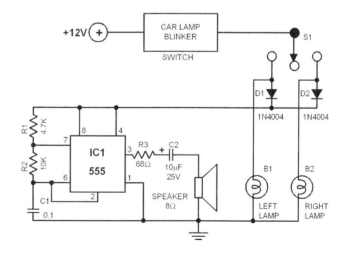

Diagram 95.0   Signal Light Clicker

Electronic Circuits - 1.0

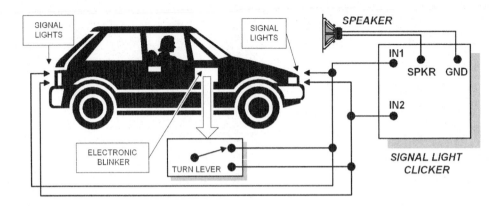

In electronic systems, however, the blinking of the car's signal light is electronically controlled and therefore no clicking sound is produced. This absence of acoustic feedback makes the driver wonder whether his signal lights were really blinking. To solve this problem, the circuit here was designed. It provides a "clicking" sound in synchronization with the blinking of the signal lights.

Instead of producing a rough click, however, it produces a pleasant tone everytime the signal light blinks. The tone's volume can be varied by using a different value for R3 (minimum = 68 ohms). You can also change the tone's frequency by changing the value of C1. The switch S in the circuit is actually the built-in signal switch in the car.

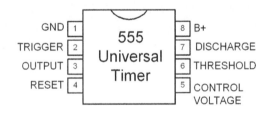

555 Universal timer IC

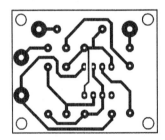

Figure 95.0 Printed Circuit Layout

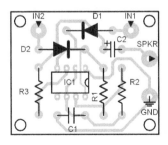

Figure 95.1 Parts Placement

Auxiliary

96 Headlamp Dimmer

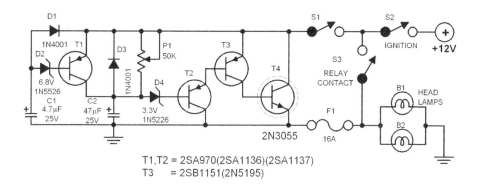

Diagram 96.0 Headlamp dimmer

Every driver knows that a high intensity light striking one's eyes can have dangerous consequencies. Cars have a dimmer switch to lower the light intensity but unfortunately, some auto headlamps are just too bright even if they are dimmed. The dimmer circuit featured here offers a better solution. When the circuit is activated, it is connected in series to the headlamps through the dimmer switch S1. The brightness of the lamps is then pulled down to a preset level which is determined by the dimmer circuit. The dim level can be preset through the potentiometer P1.

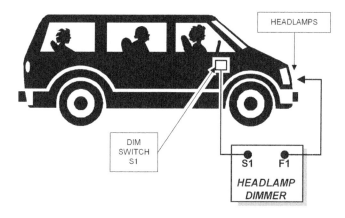

The electronic headlamp dimmer offers a more flexible control of the headlamps´ brightness and more safety.

Electronic Circuits - 1.0

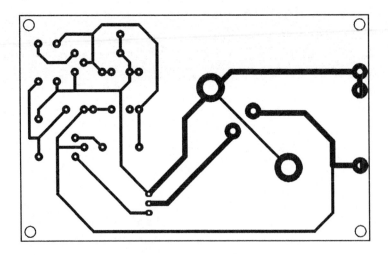

Figure 96.0 Printed Circuit Layout

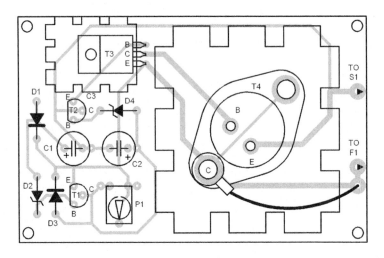

Figure 96.1 Parts Placement

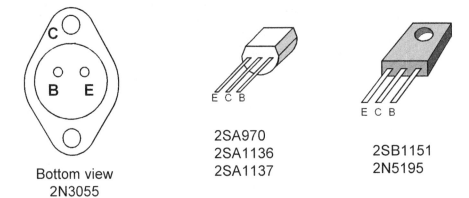

Bottom view
2N3055

2SA970
2SA1136
2SA1137

2SB1151
2N5195

Auxiliary

97 One Chip TV Audio Amp

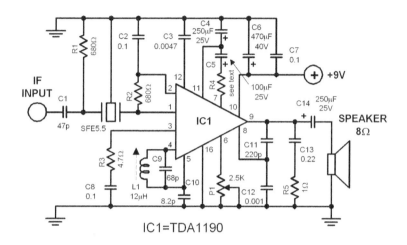

Diagram 97.0 One Chip TV Audio Amplifier

The single chip TDA 1190 integrates all important functions needed to work as a TV audio circuit. The IF signal is simply connected to its input and the output is connected to a loudspeaker. The internal functions of the IC are:

* IF amplifier and limiter
* Active lowpass
* FM demodulator
* AF preamplifier
* AF final amplifier
* Volume control

The circuit uses a minimum number of external components. It can operate from 4.5 MHz up to 6 MHz and therefore applicable for all types of TV system. The power supply can be between 12 and 24 volts. With a supply voltage of 12 volts, the power output is around 1.5 Watts (8 ohm load impedance). It delivers 4.2 Watts at 24 volts supply voltage and 16 ohms load impedance.

The input line is connected to a ceramic filter Murata SFE 5.5 MA since the input impedance is very high. The amplification of the circuit is determined by R4. The upper limit of the audio frequency range is dependent on the value of C10. The signal to noise ratio is around 70 dB while the AM suppresion is around 55 dB. Its input sensitivity is around 30μV.

Page 181

Electronic Circuits - 1.0

Table 97.0 shows the normal values of R4 according to the supply voltage and speaker load impedance. The frequency range is between 50 Hz and 12 kHz when the Vss is 24 volts and speaker impedance = 16 and R4= 18. This will become narrower between 50 Hz and 9.1 kHz when the R4 = 10 ohms. The IC must be heatsinked and the heatsink must be connected to the circuit ground.

Vss	12	24 V
Speaker impedance	8	16
R4	18	10

Table 97.0

98 Simple Electronic Organ

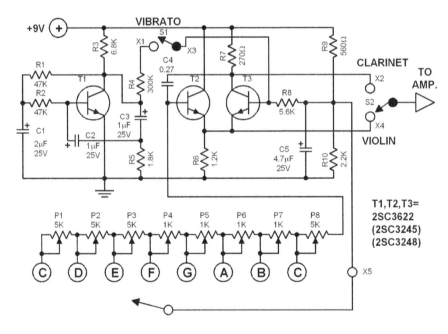

Diagram 98.0 Simple Electronic Organ

This electronic organ is very simple to construct and can provide hours of enjoyment particularly for children. The circuit is basically an emitter-coupled oscillator composed of T2 and T3. An squarewave voltage can be sampled from the collector of T3(X2). This signal gives the tone a clarinet character. Without the squarewave signal, the sound produced by the emitters of T2 and T3(X4) has a violin character.

Auxiliary

An additional vibrato signal can be added to this basic sound through switch S1. The frequency of the vibrato is approximately 6 Hz. Its amplitude is determined by the resistor R4. The value of R4 can vary from 100 up to 300K. Try experimenting with different values.

The keys can be made of either metal plates or etched printed circuit. The trimmers P1 up tp P8 adjust the pitch of each tone. The tones can be drastically changed by changing the value of C4.

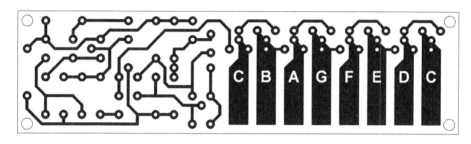

Figure 98.0 Printed Circuit Layout

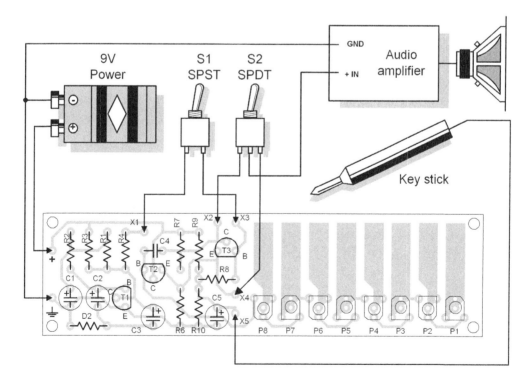

Figure 98.1 Parts Placement and external wiring layout

Page 183

99 Voltage/Frequency Converter

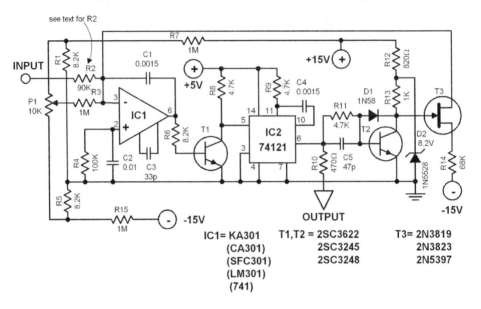

Diagram 99.0 Voltage/Frequency Converter

This voltage to frequency converter is designed to be connected to a frequency counter to display the voltage value being measured. This converter-counter combination produces a cheap but functionally complete digital voltmeter. The circuit delivers TTL compatible pulses which are 5 µS wide. The pulse frequency is proportional to the DC voltage being measured. The conversion factor is 10 kHz per volt. When the input level is 0 V the frequency output is 0 Hz.

The resistor R2 (together with C1) determines the conversion factor and has the value of around 90 K. To achieve best results use a metal film resistor for R2 in series with a multiturn trimmer resistor. This multiturn resistor will enable you to set the correct resistance value. Resistors R9 and R14 must be metal film types.

2SC3622
2SC3245
2SC3249

2N3823
2N5397

2N3819

Auxiliary

Calibration:

1. Connect the output terminals to the digital frequency counter.

2. Set the frequency counter to the appropriate range (ex. 100 Hz).

3. Short the input terminals of the converter module together.

4. Adjust P1 so that the output frequency is 0 Hz as displayed by the frequency counter (maximum allowable is 2 Hz).

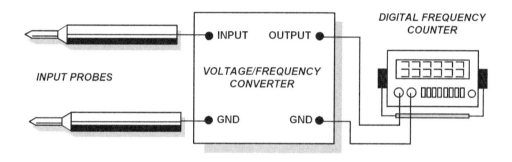

A typical combination to build a functional digital voltmeter using a digital frequency counter as the display module.

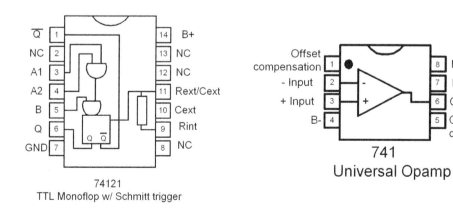

74121
TTL Monoflop w/ Schmitt trigger

741
Universal Opamp

Electronic Circuits - 1.0

100 Debounced Pulse Generator

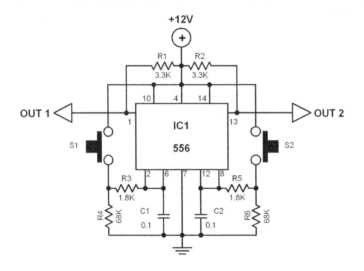

Diagram 100.0 Debounced Pulse Generator

Microprocessor systems normally require definite control signals or pulses to be able to respond correctly. Mostly, these signals are debounced first before they are introduced to processing systems. If, however, this control signal or pulse is produced with mechanical devices (like pressing a button or key), the ordinary flip-flop debounce circuits sometimes are not sufficient. It could happen that the key is released too early and the system does not recognize the control pulse. The switch might also vibrate and produce a burst of pulses which may be falsely interpreted by the processor.

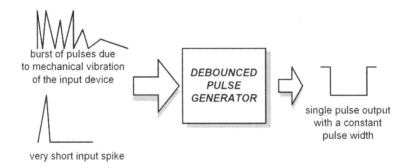

The debounced pulse generator outputs a single pulse with a constant pulse width no matter what form the input pulse (or pulse burst) has.

Auxiliary

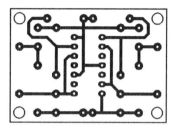
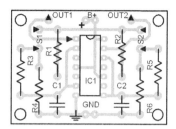

Figure 100.0 Printed Circuit Layout   **Figure 100.1** Parts Placement

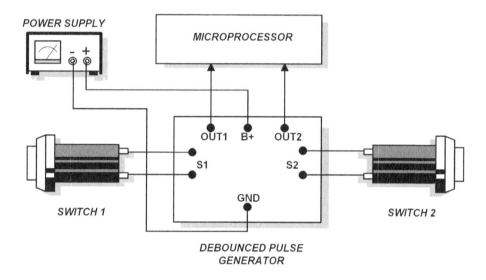

Figure 100.2 External wiring layout

The circuit presented here is a debounced pulse generator. It creates a single pulse which is independent from the length of time the input button is pressed. The pulsewidth triggered by button S1 is determined by the components R3, R4 and C1.

The components R5, R6 and C2 are responsible for shaping the pulse width triggered by button S2. Button switch S1 controls output 1 while the button S2 controls output 2. When a key or button is pressed, the corresponding output will deliver a single "0" (or low) logic pulse with a constant pulsewidth.

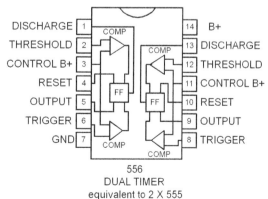

556
DUAL TIMER
equivalent to 2 X 555

Page 187

This page is intentionally blank.

APPENDICES

Specifications of the transistors used in the projects — 190

Semiconductor diode specifications — 195

Zener diode specifications — 196

Power FETs — 198

Small signal FETs — 199

Package information of FETs — 200

Three-terminal voltage regulators — 201

Package information of 3-terminal voltage regulators — 202

Printed circuit board layouts — 203

Appendices

Specifications of the transistors used in the projects

Descriptive Part of the Table:

Type
The original type designation has been taken over directly from the manufacturers, with the abbreviation of the manufacturer added in brackets only in those cases in which different manufacturers used the same type designation.

Mat.
The materials used are abbreviated as follows:
Ge	Germanium
Mos	MOS technology (metal oxide silicon)
Si	Silicon
V-MOS	Vertical MOS technology

Pol.
The polarities used are abbreviated as follows:
npn	NPN structure
n-ch	N channel type (FET)
n-p	More than one transistor with different polarities in one case
pnp	PNP structure
p-ch	P channel type (FET)

Abbreviations used in the following table:

A	Antenna amplifer	FET	Field-effect transistor
AGC	Regulating steps	FET-depl.	Field-effecttransistor, depletion type
AF	AF range	FET-enh.	Field-effect transistor, enhancement type
AM	AM range	FM	FM range
CATV	Broad band cable amplifier	fs	Fast switch
CB	CB-radio	HD	Horizontal deflection
CTV	Colour television application	hi-rel	high reliability
chop	Chopper	Idss	Drain source short-circuit current (FET)
Darl	Darlington transistor	IF	IF applications
dg	Dual Gate (FET)	in	Input stages
double	Paired types	iso	insulated
dr	Driver stages	ln	Low noise
dual	Dual transistor (differential amplifier)	min	Miniaturised version
		mix	Mixer stages
end	Final stages	nixie	Digital display tube

Appendices

osc	Oscillator stages	Ugs	Gate source voltage
pow	Power stages	UHF	UHF range > 250MHz
radiation	Aerospace applications (radiation-proof)	uni	Universal type
		Up	Pinch-off voltage (FET)
RF	RF range	VD	Vertical deflection
s	Switch	VHF	VHF range 100-250 MHz
SMP	Switch-mode power supply	Vid	Video output stages
SSB	Single side-band operation	+Diode, +di	With integrated diode
Stabi	Stabilisation	../..ns	turn-on/turn-off time
sym	Symmetrical types		
TV	Television applications		

Data Part of the Table:

In the case of the ratings, either average values are quoted (< = max.) or lower (> = min.) guaranteed values. As a rule apply at 25°C, unless otherwise indicated.

U_c

With transistors, the usual situation is for U_{CBO} (colletor base reverse bias) to be quoted, or U_{CEO} and U_{CEO} (collector emitter reverse bias). With FETs, U_{DS} (drain source voltage) is always quoted.

I_c

With transistors, I_c (collector current) is always quoted. If this is followed by (ss) in brackets, I_{CM} is quoted, i.e. the peak value of the collector current. With FETs, I_D (drain current) is always quoted.

Ptot

As a rule, the total leakage power Ptot is quoted, with RF types we always quote the RF output power P_Q, with corresponding frequency in brackets.

Amplification

The DC current gain $B(h_{FE})$ or the short-circuit current gain $ß(h_{fe})$ are always quoted as guaranteed values.

f_T

The transition frequency is always qouted in MHz.

Appendices

Specifications of the transistors used in the projects

Type	Mat.	Pol.	Description	UC [Vmax]	IC [Amax]	Ptot [Wmax]	Current Gain	fT [MHz]
MJ3001	Si	npn	Darl+diode,pow	60	10.00	150.00($25°C)	>10	
MJE243	Si	npn	AF-s-pow	100	4.00	1.50($25°C)	40.120	>40.00
MJE244	Si	npn	AF-s-pow	100	4.00	1.50($25°C)	>25	>40.00
MJE253	Si	npn	AF-s-pow	100	4.00	1.50($25°C)	40-120	>40.00
MJE4350	Si	pnp	AF-end,s-pow	100	16.00	125.00($25°C)	15	>1.00
MJE5170	Si	pnp	uni-pow	120	6.00	2.00($25°C)	15-100	>1.00
MJE5180	Si	npn	uni-pow	120	6.00	2.00($25°C)	15-100	>1.00
MPF102	Si	n-ch	FET,VHF-in,sym,mix 25V, Idss>2mA,Up<V					
MPF106	Si	n-ch	FET,VHF 25V,Idss>4mA,Up<8V					
MPS-A29	Si	npn	Darl	100	0.50	1.50($25°C)	>10	>125.00
2N708	Si	npn	s	40/15	0.20	1.20(25°C)	>15	480.00
2N1711	Si	npn	uni	75	0.50	3.00(25°C)	75	>70.00
2N1889	Si	npn	AF-s	100/60	0.50	3.00(25°C)	40-120	>50.00
2N1890	Si	npn	AF-s	100/60	0.50	3.00(25°C)	100-300	>60.00
2N1990	Si	npn	nixie	100	1.00	2.00(25°C)	>25	
2N2102	Si	npn	AF-s	120/65	1.00	5.00(25°C)	40-120	>120.00
2N2222	Si	npn	ini	0		1.80(25°C)		
2N2368	Si	npn	fs	40/15	0.20	1.20(25°C)	20-60	>400.00
2N2369	Si	npn	fs	40/15	0.20	1.20(25°C)	40-120	>500.00
2N2905	Si	pnp	uni	60/40	0.60	3.00(25°C)	100-300	>200.00
2N2904	Si	pnp	uni	60/40	0.60	3.00(25°C)	40-120	>200.00
2N3019	Si	npn	uni	140/80	1.00	5.00(25°C)	100-300	>100.00
2N3020	Si	npn	uni	140/80	1.00	5.00(25°C)	40-120	>80.00
2N3055	Si	npn	AF-s-pow	100/60	15.00	115.00($25°C)	20-70	>2.50
2N3109	Si	npn	AF-s	80/40	1.00	5.00(25°C)	100-300	>70.00
2N3110	Si	npn	AF-s	80/40	1.00	5.00(25°C)	40-120	>60.00
2N3367	Si	n-ch	FET,uni,In	40V,Idss>0.5mA,Up<2.5V				
2N3370	Si	n-ch	FET,uni,In	40V,Idss>0.1mA,Up3.2V				
2N3454	Si	n-ch	FET,uni	50V,Idss>0.05mA,Up<2.3V				
2N3819	Si	n-ch	FET,VHF,uni,sym	25V,Idss>2mA,Up<8V				
2N3823	Si	n-ch	FET,VHF,In	30V,Idss>4mA,Up<8V				
2N3903	Si	npn	uni	60/40	0.20	1.50(25°C)	50-150	>250.00
2N3904	Si	npn	uni	60/40	0.20	1.50(25°C)	100-300	>300.00
2N3905	Si	pnp	uni	40	0.20	1.50(25°C)	50-150	>200.0
2N3906	Si	pnp	uni	40	0.20	1.50(25°C)	100-300	>250.00
2N4118	Si	n-ch	FET,uni	40V,Idss>0.08mA,Up<3V				
2N5294	Si	npn	AF-s-pow	80/70	4.00	1.80($25°C)	30-120	>0.80
2N5397	Si	n-ch	FET,VHF/UHF	25V,Idss>10mA,Up<6V				
2N5398	Si	n-ch	FET,VHF/UHF	25V,Idss>5mA,Up<6V				
2N5486	Si	n-ch	FET,VHF/UHF	25V,Idss>8mA,Up<6V				
2N6038	Si	npn	Darl+diode,pow	60	4.00	1.50($25°C)	>10	>25.00
2N6039	Si	npn	Darl+diode,pow	80	4.00	1.50($25°C)	>10	>25.00
2N6283	Si	npn	Darl+diode,pow	80	20.00	160.00($25°C)	>10	>4.00
2N6284	Si	npn	Darl+diode,pow	100	20.00	160.0($25°C)	>10	>4.00
2N6412	Si	npn	AF-s-pow	60/40	4.00	15.00($25°C)	>5	>50.00
2N6414	Si	pnp	AF-s-pow	80/60	4.00	15.00($25°C)	>5	>50.00
2SA511	Si	pnp	AF/RF/s	90/80	1.50	8.00(25°)	30-150	60.00
2SA597	Si	pnp	RF-s	50/40	1.00	6.00($25°C)	10-250	400.00
2SA761	Si	pnp	uni	110	2.00	6.30($25°)	50-240	80.00
2SA970	Si	pnp	AF,In	120	0.10	0.30(25°C)	200-700	100.0

Appendices

Specifications of the transistors used in the projects

Type	Mat.	Pol.	Description	UC [Vmax]	IC [Amax]	Ptot [Wmax]	Current Gain	fT [MHz]
2SA1016	Si	pnp	uni,ln	120/100	0.05	0.40(25°)	160-960	110.00
2SA1123	Si	pnp	uni,ln	150	0.05	0.7(25°)	65-450	200.00
2SA1136	Si	pnp	AF-in,ln	120/100	0.10	0.30(25°C)	120-560	90.00
2SA1137	Si	pnp	AF-in,on	80	0.10	0.30(25°C)	120-560	90.00
2SA1141	Si	pnp	AF/Rf-pow	115	10.00	2.00($25°C)	100	80.00
2SA1285	Si	pnp	uni	120	0.20	0.90(25°C)	150-800	200.00
2SA1285A	Si	pnp	uni	150	0.10	0.90(25°C)	150-500	200.00
2SA1515	Si	pnp	uni	40/32	1.00	0.50(25°C)	82-390	150.00
2SA1705	Si	pnp	AF,s	60/50	1.00	0.90(25°C)	>30	150.00
2SA1706	Si	pnp	AF-s	60/50	2.00	1.00(25°C)	>40	150.00
2SB633	Si	pnp	AF-s-pow	100/85	6.00	40.00($25°C)	40-320	15.00
2SB764	Si	pnp	uni	60/50	1.00	0.90(25°C)	60-320	150.00
2SB822	Si	pnp	Af-dr/end	40/32	2.00	0.75(25°C)	82-390	100.00
2SB826	Si	pnp	s-pow	60/50	7.00	60.00($25°C)	>30	10.00
2SB867	Si	pnp	AF/s-pow,lo-sat	130/80	3.00	30.00($25°C)	60-260	30.00
2SB868	Si	pnp	AF/s-pow,lo-sat	130/80	4.00	35.00($25°C)	60-260	30.00
2SB869	Si	pnp	AF/s-pow,lo-sat	130/80	5.00	40.00($25°C)	60-260	30.00
2SB870	Si	pnp	AF/s-pow,lo-sat	120/80	7.00	40.00($25°C)	60-260	30.00
2SB874	Si	pnp	AF/s-pow, TV-VD	100/60	2.00	20.00($25°C)	>40	250.00
2SB909	Si	pnp	AF-dr/end	40/32	1.00	1.00(25°C)	82-390	150.00
2SB911	Si	pnp	AF-dr/end	40/32	2.00	1.00(25°C)	82-390	100.0
2SB920	Si	pnp		120/80				
2SB921	Si	pnp		120/80				
2SB1064	Si	pnp	AF-s-pow	60/50	3.00	1.50($25°)	60-320	70.00
2SB1114	Si	pnp	min,uni	20	2.00	2.00($25°C)	135-600	180.00
2SB1116	Si	pnp	uni	60/50	1.00	0.75(25°C)	135-600	120.00
2SB1142	Si	pnp	s-pow	60/50	2.50	10.00(25°C)	>35	140.00
2SB1143	Si	pnp	s-pow	60/50	4.00	10.00(25°C)	>40	150.00
2SB1144	Si	pnp	AF/s-pow,lo-sat	120/100	1.50	10.00(25°C)	>30	100.00
2SB1230	Si	pnp	AF/s-pow,lo-sat	110/100	15.00	100.00($25°C)	50-140	
2SB1231	Si	pnp	AF/s-pow,lo-sat	110/100	25.00	120.00($25°C)	50-140	
2SB1232	Si	pnp	AF/s-pow,lo-sat	110/100	40.00	150.00($25°C)	50-140	
2SC270	Si	npn	s-pow	270/75	5.00	50.00($25°C)	24-40	22.00
2SC460	Si	npn	AM-in/mix/osc	30	0.10	0.20(25°C)	35-200	230.00
2SC696	Si	npn	uni	100/60	3.00	0.75(25°C)	30-173	100.00
2SC763	Si	npn	VHF	25/12	0.02	0.10(25°C)	20-300	>400.00
2SC829	Si	npn	AM/FM-in/mix/osc	30/20	0.03	0.40(25°C)	40-500	230.00
2SC959	Si	npn	uni	120/80	0.70	0.70(25°C)	40-200	100.00
2SC1324	Si	npn	UHF-CATV	35/25	0.15	3.00(25°C)	10-35	
2SC1876	Si	npn	Darl	100/70	0.50	0.80(25°C)	>20	
2SC2124	Si	npn	TV-HD	220/800	2.00	5.00($90°C)	20	4.00
2SC2125	Si	npn	TV-HD	220/800	5.00	50.00($25°C)	8-25	5.00
2SC2270	Si	npn	lo-sat	50/20	5.00	1.00(25°C)	>70	100.00
2SC2334	Si	npn	s-pow,dc-dc conv.	150/100	7.00	40.00($25°C)	>20	
2SC2459	Si	npn	uni	120	0.10	0.20(25°C)	200-700	100.00
2SC2675	Si	npn	AF,ln	80	0.10	0.30(25°C)	180-820	120.00
2SC2724	Si	npn	FM-IF	30/25	0.03	0.20(25°C)	25-300	200.00
2SC3112	Si	npn	AF,ln	50	0.15	0.40(25°C)	600-3600	250.00
2SC3179	Si	npn	AF-pow	80/60	4.00	30.00($25°C)	100	15.00
2SC3245	Si	npn	uni	120	0.10	0.90(25°C)	150-800	200.00

Appendices

Specifications of the transistors used in the projects

Type	Mat.	Pol.	Description	UC [Vmax]	IC [Amax]	Ptot [Wmax]	Current Gain	fT [MHz]
2SC3245A	Si	npn	uni	150	0.10	0.90(25°C)	400-800	200.00
2SC3248	Si	npn	uni	180	0.10	0.90(25°C)	150	130.00
2SC3358	Si	npn	UHF	20/12	0.10	0.25(25°C)	50-300	7000.00
2SC3420	Si	npn	lo-sat	50/20	5.00	10.00(25°C)	>70	100.00
2SC3622	Si	npn	AF-s,hi-beta	60/50	0.15	0.25(25°C)	1000-3200	250.00
2SC4308	Si	npn	VHF-A	30/20	0.30	0.60(25°C)	50-200	2500.00
2SD386	Si	npn	TV-VD	200/120	3.00	1.75($25°C)	40-320	8.00
2SD406	Si	npn	Darl	100	2.00	15.00(25°C)	>2000	
2SD613	Si	npn	AF-s-pow	100/85	6.00	40.00($25°C)	40-320	15.00
2SD614	Si	npn	Darl	100/80	3.00	0.80(25°C)	3000	15.00
2SD621	Si	npn	TV_HD	2500/900	3.00	50.00($25°C)	3-15	
2SD628	Si	npn	Darl+diode,pow	100	10.00	80.00($25°C)	>1000	
2SD629	Si	npn	Darl+diode,pow	100	10.00	100.00($25°C)	>1000	
2SD688	Si	npn	Darl,pow	100	1.50	0.80($25°C)	>10	
2SD712	Si	npn	AF-s-pow	100	4.00	30.00($25°C)	55-300	8.00
2SD726	Si	npn	AF-s-pow	100/80	4.00	40.00($25°C)	35-320	10.00
2SD729	Si	npn	Darl+diode,pow	100	20.00	125.00($25°C)	>1000	
2SD781	Si	npn	s-pow,TV-HD	150/60	2.00	1.00(25°C)	150	
2SD826	Si	npn		60/20	5.00	1.00($25°C)	120-560	120.00
2SD838	Si	npn	TV-HD,s-pow	2500/900	3.00	50.00($25°C)	3-15	
2SD892A	Si	npn	Darl	60/50	0.50	0.40(25°C)	>8000	150.00
2SD1049	Si	npn	AF-s-pow	120/80	25.00	80.00($25°C)	>20	
2SD1062	Si	npn	s-pow	60/50	12.00	40.00($25°C)	>30	10.00
2SD1153	Si	npn	Darl	80750	1.50	0.90(25°C)	>40	120.00
2SD1177	Si	npn	AF-pow,TV-HD	100/60	2.00	20.00($25°C)	>40	230.00
2SD1237	Si	npn	s-pow	90/80	7.00	1.75($25°C)	>30	20.00
2SD1238	Si	npn	s-pow	90/80	12.00	80.00($25°C)	>30	20.00
2SD1639	Si	npn	AF-s-pow	100/80	2.20	10.00($25°C)	40-200	
2SD1684	Si	npn	AF/s-pow,lo-sat	120/100	1.50	10.00(25°C)	>30	120.00
2SD1685	Si	npn	AF/s-pow,lo-sat	60/20	5.00	10.00(25°C)	>95	120.00
2SD1691	Si	npn	AF-s-pow	60	5.0	20.00(25°C)	100-400	
2SD1840	Si	npn	AF/s-pow,lo-sat	110/100	15.00	100.00($25°C)	50-140	
2SD1841	Si	npn	AF/s-pow,lo-sat	110/100	25.00	120.00($25°C)	50-140	
2SD1842	Si	npn	AF/s-pow,lo-sat	110/100	40.00	150.00($25°C)	50-140	
2SD2116	Si	npn	Darl	80/50	0.70	1.00(25°C)	>40	
2SD2117	Si	npn	Darl	80/50	1.50	1.00(25°C)	>30	
2SD2213	Si	npn	Darl,AF	150/80	1.50	0.90(25°C)	>10	
2SJ165	V-MOS	p-ch	FET-enh.,	50V,0.1A,0.25W				
2SK422	V-MOS	n-ch	FET-enh.	60v,0.7A,0.9W,17/12ns				
2SK423	V_MOS	n-ch	FET-enh.	100V,0.5A,0.9W,15/20ns				
3N140	MOS	n-ch	FET-depl.,dg,FM/VHF-in	20V,Idss>5mA				
3N225	MOS	n-ch	FET-depl.,dg, UHF	25V,Idss>1mA,Up<4V				
3SK35	MOS	n-ch	FET-depl.,dg,VHF	20V,Idss>3mA,Up<4V				
3SK37	MOS	n-ch	FET-depl.,dg,VHF	20V,Idss>4mA,Up<3V				
3SK45	MOS	n-ch	FET-depl.,dg,VHF	22V,Idss>4mA,Up<3V				
3SK61	MOS	n-ch	FET-depl.,dg,VHF	20V,Idss>4mA,Up<3V				
3SK72	MOS	n-ch	FET-depl.,dg,VHF	20V,Idss>2.5mA,Up<3V				
3SK77	MOS	n-ch	FET-depl.,dg,VHF	20V,Idss>3mA,Up<2.5V				
3SK85	MOS	n-ch	FET-depl.,dg,VHF	20V,Idss>4mA,Up<3V				

SEMICONDUCTOR DIODE SPECIFICATIONS

*RFR = Rectifier, Fast Recovery

Device	Type	Material	Peak Inverse Voltage, PIV (Volts)	Average Rectified Current Forward (Reverse) IO (A) (IR(A))	Peak Surge Current, IFSM 1 sec. @ 25°C (A)	Average Forward Voltage, VF (Volts)
1N34	Signal	Germanium	60	8.5 m (15.0m)		1.0
1N34A	Signal	Germanium	60	5.0 m (30.0m)		1.0
1N67A	Signal	Germanium	100	4.0 m (5.0m)		1.0
1N191	Signal	Germanium	90	5.0 m	1.0	
1N270	Signal	Germanium	80	0.2 (100 m)		1.0
1N914	Fast Switch	Silicon (Si)	75	75.0 m (25.0 n)	0.5	1.0
1N1184	RFR	Si	100	35 (10 m)		1.7
1N2071	RFR	Si	600	0.75 (10.0m)		0.6
1N3666	Signal	Germanium	80	0.2 (25.0m)		1.0
1N4001	RFR	Si	50	1.0 (0.03 m)		1.1
1N4002	RFR	Si	100	1.0 (0.03 m)		
1N4003	RFR	Si	200	1.0 (0.03 m)		1.1
1N4004	RFR	Si	400	1.0 [0.03 m)		1.1
1N4005	RFR	Si	600	1.0 (0.03 m)		1.1
1N4006	RFR	Si	800	1.0 (0.03 m)		1.1
1N4007	RFR	Si	1000	1.0 (0.03 m)		1.1
1N4148	Signal	Si	75	10.0 m (25.0 n)		1.0
1N4149	Signal	Si	75	10.0 m (25.0 n)		1.0
1N4152	Fast Switch	Si	40	20.0 m (0.05m)		0.8
1N4445	Signal	Si	100	0.1 (50.0 n)		1.0
1N5400	RFR	Si	50	3.0	200	
1N5401	RFR	Si	100	3.0	200	
1N5402	RFR	Si	200	3.0	200	
1N5403	RFR	Si	300	3.0	200	
1N5404	RFR	Si	400	3.0	200	
1N5405	RFR	Si	500	3.0	200	
1N5406	RFR	Si	600	3.0	200	
1N5767	Signal	Si		0.1 (1.0µ)		1.0
ECG5863	RFR	Si	600	6	150	0.9

*RFR = Rectifier, Fast Recovery

Appendices

ZENER DIODES SPECIFICATIONS

Zener Voltage (Volts)	Power (Watts)							
	0.25	0.4	0.5	1.0	1.5	5.0	10.0	50.0
1.8	1N4614							
2.0	1N4615							
2.2	1N4616							
2.4	1N4617	1N4370,A	1N4370,A,1N5221,B 1N5985,B					
2.5			1N5222B					
2.6	1N702,A							
2.7	1N4618	1N4371,A	1N4371,A,1N5223,B 1N5839, 1N5986					
2.8			1N5224B					
3.0	1N4619	1N4372,A	1N4372,1N5225,B 1N5987					
3.3	1N4620	1N746,A 1N764 A 1N5518	1N746A 1N5226,B 1N5988	1N3821 1N4728,A	1N5913	1N5333,B		
3.6	1N4621	1N747,A 1N5519	1N747A 1N5227,B,1N5989	1N3822 1N4729,A	1N5914	1N5334,B		
3.9	1N4622	1N748,A 1N5520	1N748A,1N5228,B 1N5844, 1N5990	1N3823 1N4730,A	1N5915	1N5335,B	1N3993A	1N4549,B 1N4557,B
4.1	1N704,A							
4.3	1N4623	1N749,A 1N5521	1N749,A 1N5229,B 1N5845,1N5991	1N3824 1N4731,A	1N5916	1N5336,B	1N3994,A	1N4550,B 1N4558,B
4.7	1N4624	1N750,A 1N5522	1N750A,1N5230,B 1N5846, 1N5992	1N3825 1N4732,A	1N5917	1N5337,B	1N3995,A	1N4551,B 1N4559,B
5.1	1N4625 1N4689	1N751 A 1N5523	1N751A, 1N5231,B 1N5847,1N5993	1N3826 1N4733	1N5918	1N5338,B 1N4560,B	1N3996,A	1N4552,B
5.6	1N708A 1N4626	1N752,A 1N5524	1N752,A,1N5232,B 1N5848, 1N5994	1N3827 1N4734,A	1N5919	1N5339,B 1N4561,B	1N3997,A	1N4553,B
5.8	1N706A	1N762						
6.0				1N5233B 1N5849			1N5340,B	
6.2	1N709,1N4627 MZ605,MZ610 MZ620,MZ640	1N753,A 1N821,3,5, 7,9; A	1N753,A 1N5234,B, 1N5850 1N5995	1N3828,A 1N4735,A	1N5920	1N5341,B 1N4562,B	1N3998,A	1N4554,B
6.4	1N4565-84,A							
6.8	1N4099	1N754,A 1N957,B 1N5526	1N754,A 1N757,B 1N5235,B 1N5851 1N5996	1N3016,B 1N3829 1N4736,A	1N3785 1N5921	1N5342,B	1N2970,B 1N3999,A	1N2804B 1N3305B 1N4555, 1N4563
7.5	1N4100	1N755,A 1N958,B 1N5527	1N755A,1N958,B 1N5236,B, 1N5862 1N5997	1N3017,A,B 1N3830 1N4737,A	1N3786 1N5922	1N5343,B 1N4000,A 1N4556,	1N2971,B 1N3306,B	1N2805,B 1N4564
8.0	1N707A							
8.2	1N712A 1N4101	1N756,A 1N959,B 1N5528	1N756,A 1N959,B,1N5237,B 1N5853 ,1N5998	1N3018,B 1N4738,A	1N3787 1N5923	1N5344,B	1N2972,B	1N2806,B 1N3307,B
8.4			1N3154-57,A 1N3155-57	1N3154,A				
8.5	1N4775-84,A		1N5238,B,1N5854					
8.7	1N4102					1N5345,B		
8.8		1N 764						
9.0		1N764A	1N935-9;A,B					

ZENER DIODES SPECIFICATIONS

Zener Voltage (Volts)	\	\	\	Power (Watts)	\	\	\	\
	0.25	0.4	0.5	1.0	1.5	5.0	10.0	50.0
9.1	1N4103	1N757,A 1N960,B 1N5529	1N757,A, 1N960,B 1N5239,B, 1N5855 1N5999	1N3019,B 1N4739,A	1N3788 1N5924	1N5346,B	1N2973,B	1N2807,B 1N3308,B
10.0	1N4104	1N758,A 1N961,B 1N5530,B	1N758,A, 1N961,B 1N5240,B, 1N5856 1N6000	1N3020,B 1N4740	1N3789 1N5925	1N5347,B	1N2974,B	1N2808,B 1N3309,A,B
11.0	1N715,A 1N4105	1N962,B 1N5531	1N962,B,1N5241,B 1N5857, 1N6001	1N3021,B 1N4741,A	1N3790 1N5926	1N5348,B	1N2975,B	1N2809,B 1N3310,B
11.7	1N716,A 1N4106		1N941,A,B					
12.0		1N759,A 1N963,B 1N5532	1N759,A ,1N963,B 1N5242,B, 1N5858 1N6002	1N3022,B 1N4742,A	1N3791 1N5927	1N5349,B	1N2976,B	1N2810,B 1N3311,B
13.0	1N4107	1N964,B 1N5533	1N964,B,1N5243,B 1N5859,1N6003	1N3023,B 1N4743,A	1N3792 1N5928	1N5350,B	1N2977,B	1N2811,B 1N3312,B
14.0	1N4108	1N5534	1N5244B, 1N5860			1N5351,B	1N2978,B	1N2812,B 1N3313,B
15.0	1N4109	1N965,B 1N5535	1N965,B,1N5245,B 1N5861,1N6004	1N3024,B 1N4744A	1N3793 1N5929	1N5352,B	1N2979,A,B	1N2813,A,B 1N3314,B
16.0	1N4110	1N966,B 1N553,B	1N966,B,1N5246,B 1N5862, 1N6005	1N3025,B 1N4745,A	1N3794 1N5930	1N5353,B	1N2980,B	1N2814,B 1N3315,B
17.0	1N4111	1N5537	1N5247,B 1N5863			1N5354,B	1N2981B	1N2815,B 1N3316,B
18.0	1N4112	1N967,B 1N5538	1N967,B 1N5248,B 1N5864, 1N6006	1N3026,B 1N4746,A	1N3795 1N5931	1N5355,B	1N2982,B	1N2816,B 1N3917,B
19.0	1N4113	1N5539	1N5249,B 1N5865			1N5356,B	1N2983,B	1N2817,B 1N3318,B
20.0	1N4114	1N968,B 1N5540	1N968,B,1N5250,B 1N5866, 1N6007	1N3027,B 1N4747,A	1N3796 1N5932,A,B	1N5357,B	1N2984,B	1N2818,B 1N3319,B
22.0	1N4115	1N959,B 1N5541	1N969,B,1N5241,B 1N5867, 1N6008	1N3028,B 1N4748,A	1N3797 1N5933	1N5358,B	1N2985,B	1N2819,B 1N3320,A,B
24.0	1N4116	1N5542 1N9701B	1N970,B,1N5252,B 1N586,1N6009	1N3029,B 1N4749,A	1N3798 1N5934	1N5359,B	1N2986,B	1N2820,B 1N3321,B
25.0	1N4117	1N5543	1N5253,B 1N5869			1N5360,B	1N2987B	1N2821,B 1N3322,B
27.0	1N4118	1N971,B	1N971,1N5254,B 1N5870,1N6010	1N3030,B 1N4750,A	1N3799 1N5935	1N5361,B	1N2988,B	1N2822B 1N3323,B
28.0	1N4119	1N5544	1N5255,B,1N5871			1N5362,B		
30.0	1N4120	1N972,B 1N5546	1N972,B,1N5256,B 1N5872,1N6011	1N3031,B 1N4751,A	1N3800 1N5936	1N5363,B	1N2989,B	1N2823,B 1N3324,B
33.0	1N4121	1N973,B 1N5546	1N973,B,1N5257,B 1N5873,1N6012	1N3032,B 1N4752,A	1N3801 1N5937	1N5364,B	1N2990,A,B	1N2824,B 1N3325,B
36.0	1N4122	1N974,B	1N974,B,1N5258,B 1N5874,1N6013	1N3033,B 1N4753,A	1N3802 1N5938	1N5365,B	1N2991,B	1N2825,B 1N3326,B
39.0	1N4123	1N975,B	1N975,B, 1N5259,B 1N5875 ,1N6014	1N3034,B 1N4754,A	1N3803 1N5939	1N5366,B	1N2992,B	1N2826,B 1N3327,B
43.0	1N4124	1N976,B	1N976,B,1N5260,B 1N5876,1N6015 1N2994B	1N3035,B 1N4755,A 1N2828B 1N3329B	1N3804 1N5940	1N5367,B	1N2993,A,B	1N2827,B 1N3328,B
45.0								

Appendices

POWER FETs

Device No.	Type	Max. Diss. (W)	Max. V_{DS} (Volts)	Max I_D (A)#	Gfs mmhos (typ.)	Input C_{iss} (pF)	Output C_{oss} (pF)	Approx. Upper Freq. (MHz)	Case	Pack-Type Mnfr.	General applications age/
DV1202S	N-Chan.	10	50	0.5	100k	14	20	500	.380 SOE	1/S	RF power amp., oscillator
DV1202W	N-Chan.	10	50	0.5	100k	14	20	500	C-220	5/S	RF power amp., oscillator
DV1205S	N-Chan.	20	50	1	200k	26	38	500	.380 SOE	1/S	RF power amp., oscillator
DV1205W	N-Chan.	20	50	1	200k	26	98	500	C-220	5/S	RF power amp., oscillator
2SK133	N-Chan.	100	120	7	1M	600	350	1	TO-3	6/H	AF pwr. amp., switch (complem to 2SJ48)
2SK134	N-Chan.	100	140	7	1M	600	350	1	TO-3	6/H	AF pwr. amp., switch (complem to 2SJ49)
2SK135	N-Chan.	100	160	7	1M	600	350	1	TO-3	6/H	AF pwr. amp., switch (complem to 2SJ50)
2SJ48	P-Chan.	100	120	7	1M	900	400	1	TO-3	6/H	AF pwr. amp., switch (complem to 2SK133)
2SJ49	P-Chan.	100	140	7	1M	900	400	1	TO-3	6/H	AF pwr. amp., switch (complem to 2SK134)
2SJ50	P-chan.	100	160	7	1M	900	400	1	TO-3	6/H	AF pwr. amp., switch (complem to 2SK135)
VMP4	N-Chan.	25	60	2	170K	32	4.8	200	.380 SOE	1/S	VHF pwr. amp., rcvr front end (rf amp., mixer).
VN10KM	N-Chan.	1	60	0.5	100K	48	16	*	TO-92	2/S	High-speed line driver, relay driver, LED stroke driver
VN64GA	N-Chan.	80	60	12.5	150K	700	325	30	TO-3	3/S	Linear amp., power-supply switch, motor control
VN66AF	N-Chan.	15	60	2	150K	50	50	*	TO-202	4/S	High-speed switch, HF linear amp., audio amp. line driver.
VN66AK	N-Chan.	8.3	60	2	250K	93	6	100	TO-39	7/S	RF pwr. amp., high-current analog switching
VN67AJ	N-Chan.	25	60	2	250K	33	7	100	TO-3	3/S	RF pwr.amp.,high-current switching
VN89AA	N-Chan.	25	80	2	250K	50	10	100	TO-3	3/S	High-speed switching, HF linear amps., line drivers.
IRF100	N-Chan.	125	80	16	300K	900	25	*	TO-3	3/S	High-speed switching, audio inverters.
IRF101	N-Chan.	125	60	16	300K	900	25	*	TO-3	3/S	Same as IRF100

Legend: #25°C (case) S = M/A-COM H = Hitachi IR = International Rectifier. Mnfr = Manufacturer

SMALL-SIGNAL FETs

Device No.	Type	Max. Diss. (mW)	Max. V_{DS} (Volts)	Max I_D	Min G_{fs} (mA)# (mS)	Input C (pF)	$V_{GS(off)}$	Upper Freq. (volts)(MHz)	Noise Figure (MHz)	Case Type (typ)	/Mnfr.	General applications
2N4416	N-JFET	300	30	-15	4.5K	4	-6	450	400 MHz 4 dB	TO-72	1/S,M	VHF/UHF/RF amp.mix., osc.
2N5484	N-JFET	310	25	30	2.5K	5	-3	200	200 MHz 4 dB	TO-92	2/M	VHF/UHFamp,mix., osc.
2N5485	N-JFET	310	25	30	3.5K	5	-4	400	400 MHz 4 dB	TO-92	2/S	VHF/UHF/RF amp.mix., osc.
3N200	N-Dual-Gate MOSFET	330	20	50	10K	4-8.5	-6	500	400 MHz 4.5 dB	TO-72	3/R	VHF/UHF/RF amp.mix., osc.
3N202	N-Dual-Gate MOSFET	360	25	50	8K	6	-5	200	200 MHz 4.5 dB	TO-72	3/S	VHF amp., mixer
MPF102	N-JFET	310	25	20	2K	4.5	-8	200		TO-92	2/N,M	HF/VHF amp.,mix., osc.,
MPF106/ 2N5484	N-JFET	310	25	30	2.5K	5	-6	400	200 MHz 4 dB	TO-92	2/N,M	HF/VHF/UHF amp.,mix.,osc.
40673	N-Dual-Gate MOSFET	330	20	50	12K	6	-4	400	200 MHz 6 dB	TO-72	3/R	HF/VHF/UHF amp. mix., osc.
U300	P-JFET	300	-40	20	8K	-50	+10	*	400 MHz	TO-18	4/S	General Purpose amp.
U304	P-JFET	350	-30	-50	10K	27	+10	*	*	TO-18	4/S	analog switch, chopper
U310	N-JFET	500	30	60	10K	2.5	-6	450	450 MHz 3.2 dB	TO-52	5/S	common-gate VHF/UHF amp.,osc., mixer
300		30										
U350	N-JFET Quad	1W	25	60	9K	5	-6	100	100 MHz 7 dB	TO-99	6/S	matched JFET doubly bal. mixer
U431	N-JFET Dual	300	25	30	10K	5	-6	100	*	TO-99	7/S	matched JFET cascade amp., balanced mixer

#25°C   S = Siliconix Inc.   R = RCA   N = National Semiconductor   M = Motorola

Appendices

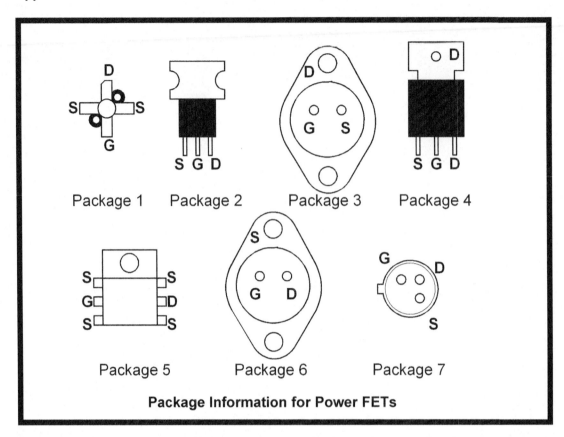

Package Information for Power FETs

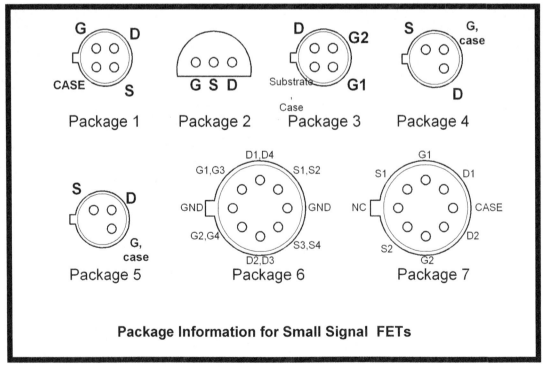

Package Information for Small Signal FETs

Appendices

Three-Terminal Voltage Regulators

!! Listed numerically by device

Device	Description	Voltage	Current (Amps)	Package
317	Adj. Pos	+1.2 to +37	0.5	TO-205
317	Adj. Pos	+1.2 to +37	1.5	TO-204,TO-220
317L	Low Current Adj. Pos	+1.2 to +37	0.1	TO-205,TO-92
317M	Med Current Adj. Pos	+1.2 to +37	0.5	TO-220
350	High Current Adj. Pos	+1.2 to +33	3.0	TO-204,TO-220
337	Adj. Neg	-1.2 to -37	0.5	TO-205
337	Adj. Neg	-1.2 to -37	1.5	TO-204,TO-220
337M	Med Current Adj. Neg	-1.2 to -37	0.5	TO-220
309		+5	0.2	TO-205
309		+5	1.0	TO-204
323		+5	9.0	TO-204,TO-220
140-XX	Fixed Pos	Note #	1.0	TO-204,TO-220
340-XX			1.0	TO-204,TO-220
78XX			1.0	TO-204,TO-220
78LXX			0.1	TO-205,TO-92
78MXX			0.5	TO-220
78TXX			3.0	TO-204
79XX	Fixed Neg	Note #	1.0	TO-204,TO-220
79LXX			0.1	TO-205,TO-92
79MXX			0.5	TO-220

Legend:

Adj.	= Adjustable
Med	= Medium
Neg	= Negative
Pos	= Positive

Note # - XX indicates the regulated voltage; which may be anywhere from 1.2 volts to 35 volts. For example a 7808 is a positive 8-volt regulator, and a 7912 is a negative 12-volt regulator.

The regulator package may be denoted by an additional suffix, according to the following:

Package	Suffix
TO-204 (TO-3)	K
TO-220	T
TO-205 (TO-39)	H,G
TO-92	P,Z

Example:
A 7815K is a positive 15-volt regulator in a TO-204 package. An LM340T-8 is a positive 8-volt regulator in a TO-220 package. In addition, different manufacturers use different prefixes. An LM7812 is equivalent to a µA 7812 or MC7812.

Appendices

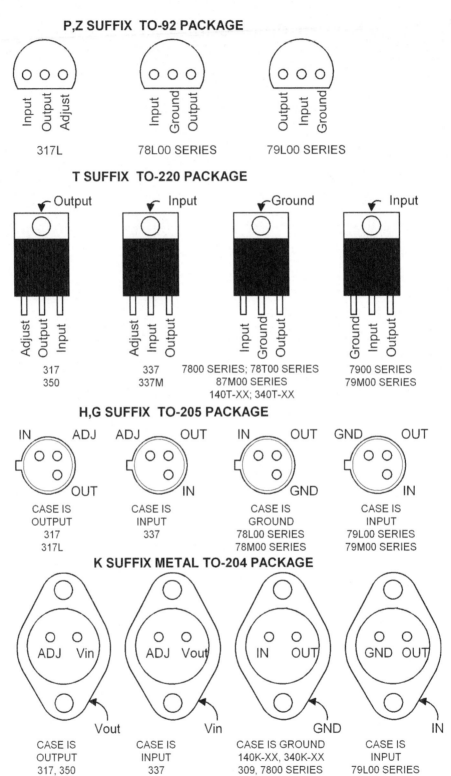

Appendices

PRINTED CIRCUIT BOARD LAYOUTS

All printed circuit board layouts in this collection are once again printed in the following pages. You can either cut out or photocopy these pages to make a separate file for quick reference.

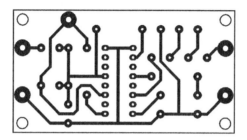

page 13 single IC 2.5W amplifier

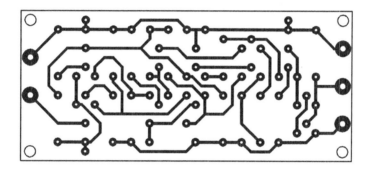

page 15 Audio Peak Meter

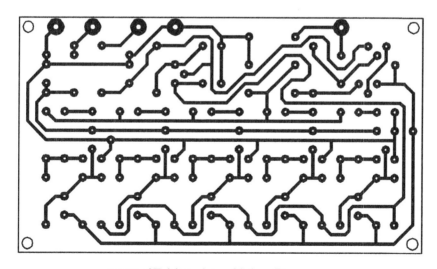

page 17 Very Low Noise Preamp

Appendices

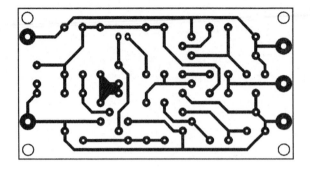

page 19 Signal Clip Indicator

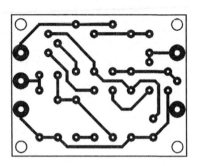

page 21 Dyna Audio Compressor

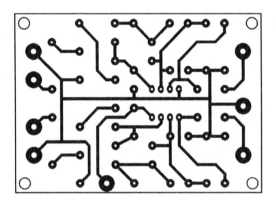

page 23 Stereo Mixer Input Preamp module

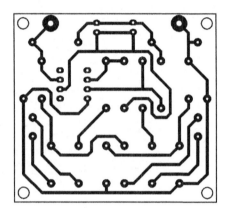

page 26 Channel Balance Indicator

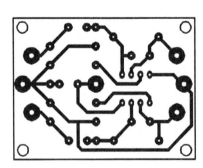

page 23 Stereo Mixer main mixer module

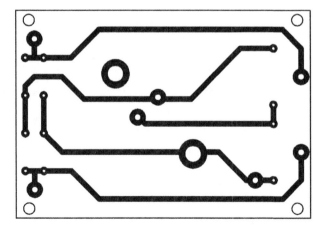

page 28 Tweeter Guardian

Appendices

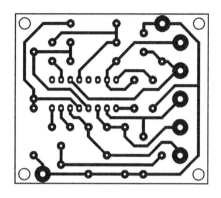

page 30 Low Noise Preamp

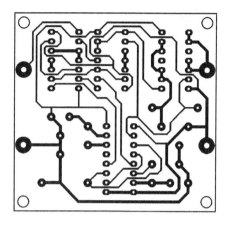

page 37 Digital Bandpass Filter

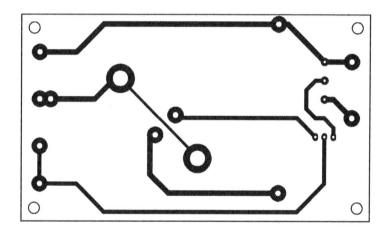

page 41 Adjustable Dummy Load

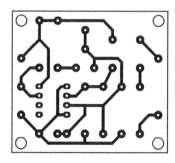

page 42 Optocoupler

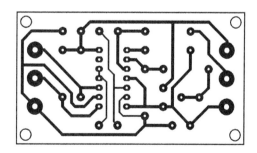

page 44 Voltage/Frequency Converter

Appendices

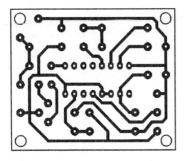

page 47 Music in a chip

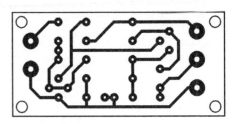

page 51 Atmospheric Disturbance Detector

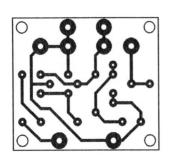

page 78 VHF Dip Meter

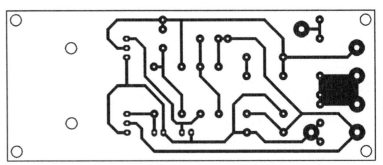

page 69 DC Flourescent Lamp

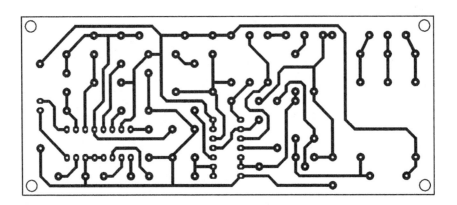

page 68 Infrared Switch receiver module

Appendices

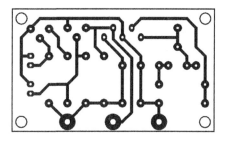

page 71 Thermonitor

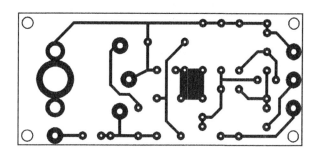

page 76 Diode AM Receiver

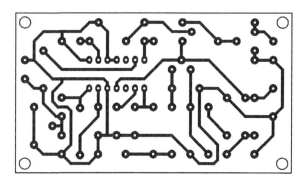

page 77 SSB from SW Adapter

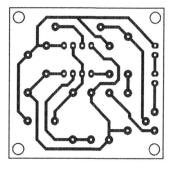

page 66 Infrared Switch transmitter module

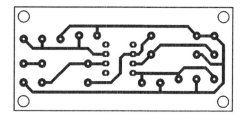

page 82 Active Impedance Converter 50-50 ohms

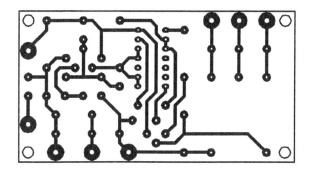

page 87
Voice Operated Switch

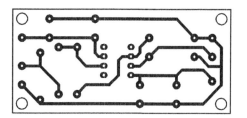

page 83 Active Impedance Converter 75-50 ohms

Appendices

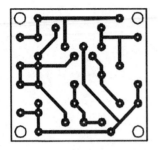

page 90 Alternating Lamps

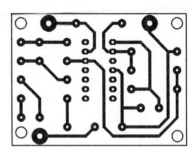

page 93 Projector Film Changer encoder module

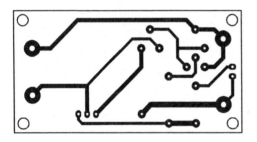

page 99 Polarity Protected Charger

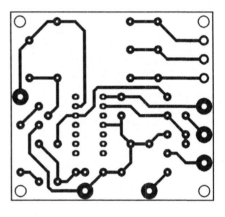

page 94 Projector Film Changer decoder module

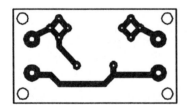

page 100 Overvoltage Crowbar circuit 1

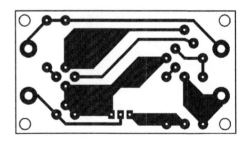

page 103 Power Supply Regulator

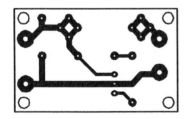

page 101 Overvoltage Crowbar circuit 2

Appendices

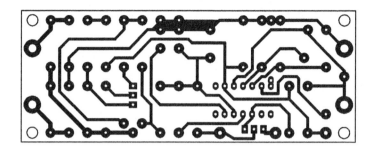

page 105 PS with Dissipation Limiter

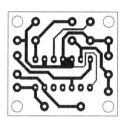

page 107 DC to DC Converter

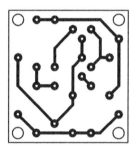

page 106 Stable Z-Voltage Source

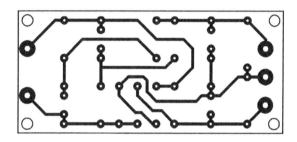

page 110 Symmetrical Auxiliary PS

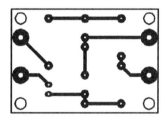

page 115 Diode Tester

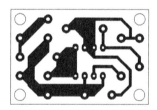

page 108 Versatile Power Supply

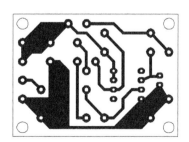

page 112 Low Drop Regulator

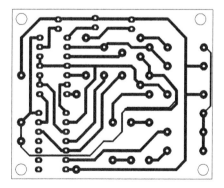

page 117 Logic Probe

Appendices

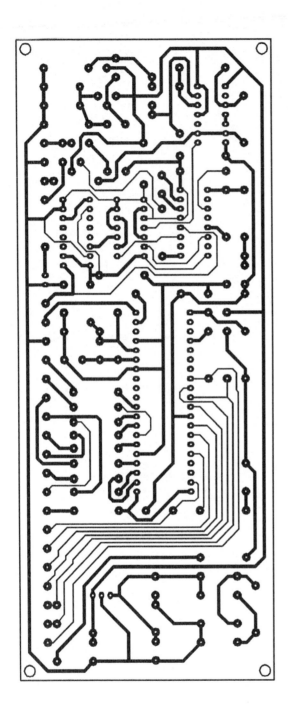

page 124 Digital AF Counter

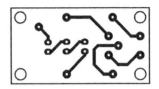

page 126 LED Current Source circuit 1

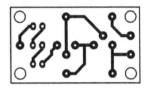

page 126 LED Current Source circuit 2

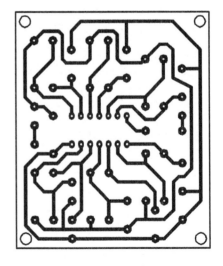

page 128 Wideband Signal Injector

Appendices

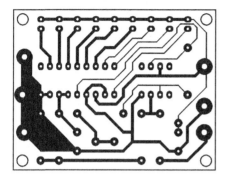

page 130 Tendency Indicator

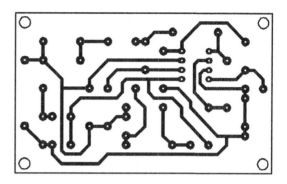

page 132 Wienbridge Oscillator

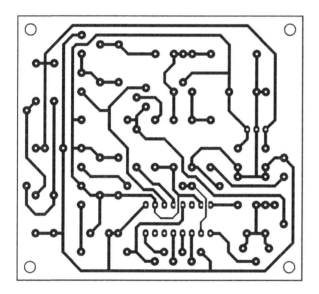

page 138 B&W TV Pattern Generator

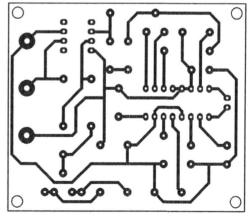

page 145 Infrared Interface transmitter module

page 146 Infrared Interface receiver module

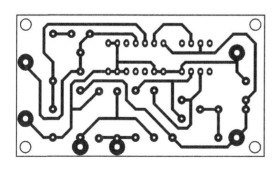

page 140 Acoustic Continuity Tester

Appendices

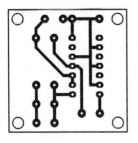

page 150 Flipflop with Inverters

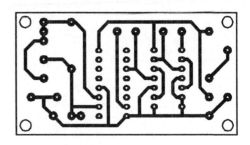

page 159
Crystal Controlled Timebase

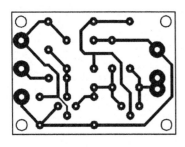

page 161
48 MHz Oscillator

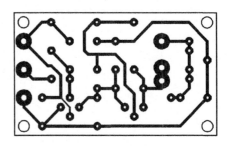

page 162 48 MHz Oscillator w/ LC

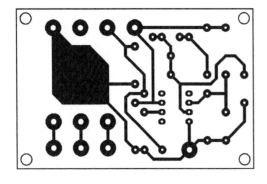

page 167 Light Activated Switch

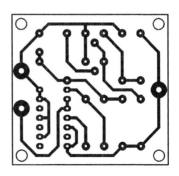

page 169 Automatic Resetter

Appendices

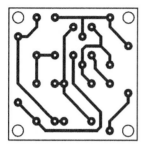

page 171 DC Voltage Doubler

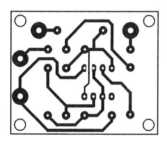

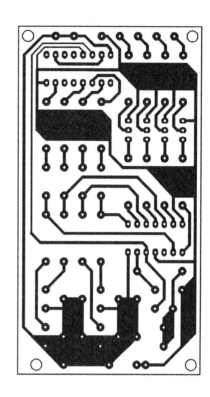

page 174 Running Light

page 178 Signal Light Clicker

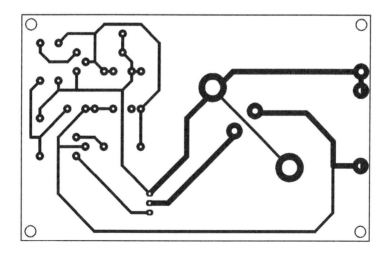

page 180 Headlamp Dimmer

Appendices

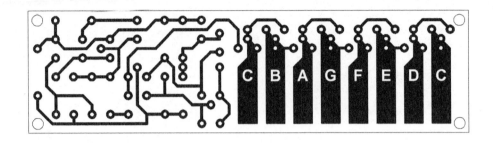

page 183 Simple Electronic Organ

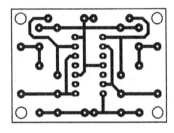

page 187
Debounced Pulse Generator

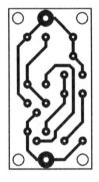

page 176
Adjustable Zener Diode

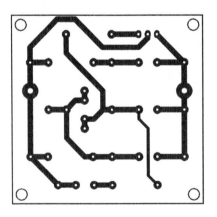

page 54
Lamp Dimmer/Speed Regulator

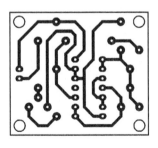

page 58
Optically Coded Key

Index

A

acoustic feedback 177
alarm 64
AM demodulator 31
AM Receiver 76
amplification 29
amplification factor 16, 175
amplifier 12, 50, 190
amplitude 183
analog CMOS switch 24
analog switch 31
antenna 50, 81, 190
astable multivibrator 127
audio amplifier 77
audio compressor 20
audio processor 32
audio signal 20
automatic answering machine 21
automatic volume control 29

B

balance 25
balance analyzer 25
bandpass 35
battery charger 42, 98
baud 148
BCD 62
BFO 48
bipolar 95
bridge rectifier 110
buffer 34
bus driver 153

C

carrier frequency 31
cascaded transistor 39
CGA 153
charge monitor 173
choke coil 69
clipping 18
clock generator 24, 96, 173
clock signal 31, 84

CMOS 156
coil microphone 29
comparator 42
compressor 20, 32
Constant Current 100, 101
Controller 95
conversion factor 44, 184
converter 80, 119, 133, 135, 164
counter 122, 133, 184
Crowbar 100
crystal 161
Current Monitored Supply 107
current sink 40
current source 126

D

darlington pair 27
Data Monitor 151
DC Converter 107
debounce 186
decimal divider 157
decoder 92, 142
demodulated signal 31
detector 50, 73
Digital Clock 142, 143
Digital Panel Meter 131
digital timer 62
dimmer 54, 179
diode 31, 76, 114
diode junction 38
diode receiver 76
dip meter 78
DIP switch 62
dissipation 104
distortion factor 16
DMM 133
DONT CARE 116
Double Alarm 57, 58
dummy load 40
duty cycle 24, 157
dynamic microphone 29

Index

E

EGA 153
electric motor 55
electronic fuse 52
electronic organ 182
electronic pushbutton 43
EPROM 107

F

feedback technique 20
FET 39, 102
Filter 24, 79
filter 36, 48, 77, 80
flip-flop 84, 150, 160, 186
flourescent lamp 69
frequency converter 44, 135
frequency counter 122, 133, 184
function generator 156
fuse 52

G

galvanic isolation 60
generator 136, 156, 157, 161, 163, 168

H

Hercules 153
HF 162
HF type transistor 34
HIFI 33
hybrid transistor 39
hybrid zener diode 176

I

IF signal 181
impedance 82
impedance converter 82
infrared 57, 66, 144
infrared beam 66
infrared detector 44, 45
initial load 110
injector 127
input impedance 34, 190
integrator 172

intercom 32, 56
Interface Monitor 146
intermodulation interference 32
isolation 175

L

LC circuit 78
LDR 64, 166
LED 71, 126, 174
LED Opto-Coupler 38
LED Reference Diode 40
limiter 32
LM1801 64
LM335 119
load current 40
load resistance 40
logic analyzer 118
logic probe 116
low pass 48
low pass filter 31, 77
low wave station 81

M

MC3479 96
metal film 16, 184
microphone 16, 32, 56
Microphone Processor 32
Mini Amplifier 14
mixer 22, 33
MK50398 121
monitor 71, 151, 152
monoflop 36
Morse Code 79
morse code filter 80

N

N-channel 52
noise level 16
noise problem 33
noise signal 168

Index

O

opamp 88, 166, 172
OPL100 60
optical key 57
optocoupler 42, 91, 174, 175
oscillation 135
oscillator 84, 131, 159
overheating 13
overload peak 27
Overvoltage 100

P

parabolic 16
pattern generator 136
peak amplitude 171
peak level 14
peak value 15
phase error 31
photodiode 60
phototransistor 42, 57
power dissipation 40
power output 12
power supply 102, 108, 109
Power Zener 172
preamp 18, 23, 29, 56
probes 139
PTC 27
pulse frequency 184
pulse generator 157, 187
pulse processing 122

R

ratio 121
receiver 67, 76, 136, 144, 190
rectifier 14, 110
reference current 39
reference voltage 88
regulator 102, 108, 111
relay 86, 88, 174
repeater 86
reset 169
RGB 151
rotary switch 91

RS232 147
RTTY Converter 78
Running Light 167, 168

S

sawtooth 172
schmitt-trigger 163
SCR 101
selector switch 65
sensor 48, 60, 64, 67, 188
shortwave radio 77
signal loss 190
Signal to noise ratio 34
silicon chip 38
sinewave 164
Sinewave Oscillator 159
smoke alarm 64
Solar Cell 38
sound effects 168
squarewave 44, 164, 182
SSB 77
SSM2016 29
Stable Power Supply 98
stator current 96
stepdown transformer 70
Stepmotor 95
Stereo 22, 25
Stereo Indicator 25
stereo sources 22
Supply for Opamps 109
SW 77, 81

T

TCA160 13
TDA1190 181
telemetry 45, 135
telephone devices 21
telephone pickup 73
televison 136
temperature 119, 188
Temperature Monitor
   71, 175, 177, 181, 182, 184, 186, 188

Index

tester 139
TFA1001W 135
thermoelement 88
thermometer 59, 188
thermonitor 71
threshold level 166
time code 62
timebase 122, 159, 161
TL496CP 108
tone code 92
tone encoder 92
transconductance 164
transducer 48
transmitter 57, 67, 86, 144
triac 54, 62, 64, 104
trianglewave 14
trimmer 183, 184
TSC9402 44
TTL 96, 184
TV 181
tweeter 27

U

UART 147
ultrasound 48

V

Variable Zener Diode 39
varicap 84
versatile timer 48
VFO 84
VGA 153

V

VHF 78
vibrato 183
video 27, 80, 81
video amplifier 27, 80, 81
video interface 158
VLF 80
voice level regulator 20
voice modulated 20
voltage doubler 171
voltage monitor 133
voltage regulator 177
voltage stabilization 39
VOX 86

W

wideband 127
wienbridge 131
wireless telemetry 45

Z

zener diode 19, 59, 100, 106 176
zener voltage 106

Printed in Great Britain
by Amazon